S0-BPI-934

After the Oil Price Collapse

After the Oil Price Collapse

OPEC, the United States, and the World Oil Market

EDITED BY WILFRID L. KOHL

The Johns Hopkins University Press
Baltimore and London

Printed in the United States of America

The Johns Hopkins University Press, 701 West 40th Street,
Baltimore, Maryland 21211
The Johns Hopkins Press Ltd., London

∞ The paper used in this book meets the minimum requirements
of American National Standard for Information Sciences—Permanence
of Paper for Printed Library Materials, ANSI Z39.48-1984.

Library of Congress Cataloging-in-Publication Data
After the oil price collapse: OPEC, the United States, and the
world oil market/edited by Wilfrid L. Kohl.
p. cm.
"February 1990."
Includes bibliographical references and index.
ISBN 0-8018-4097-X (alk. paper)
1. Organization of Petroleum Exporting Countries. 2. Petroleum
products—Prices. 3. Petroleum industry and trade. 4. Petroleum
industry and trade—United States. I. Kohl, Wilfrid L.
HD9560.1.066A63 1991
382′.42282′0601—dc20 90-46169

Contents

Figures

Tables

Foreword

AN analyst of world oil looks back on the decade of the 1980s and remembers ten years of erratic, sometimes very large price movements, the increasing importance of key suppliers in the Persian Gulf and their political and security interrelationships, the Gulf war, and, for part of the decade, an alarming reduction of oil demand among the industrial democracies. OPEC appeared to be incapable of exerting discipline on its members. Avoidance of quotas designed to limit supply and sustain prices was widespread. Partly in response to uncertainties in oil markets, several of the largest suppliers of oil embarked on downstream investments in the more important consuming markets. The reintegration of the oil industry seemed to have begun. While oil prices began to stabilize toward the end of the decade, they did so because of a larger than expected increase in oil demand, a significant rise in U.S. imports, and the inability of non-OPEC sources of oil to increase their output sufficiently to avoid a greater call on OPEC supplies. Having gone through these changes over ten years, the question now is whether the decade of the 1990s will be more of the same.

New techniques and participants in purchasing oil guarantee not volatility but short-term sensitivities to changes affecting markets. A natural desire of producers to reduce instabilities led them to formulate regional-based pricing reminiscent more of the 1950s and 1960s than of the 1970s to the mid-1980s, the years of great changes in the international oil system. At the same time, it is expected that with greater "oil power" concentrated in the Gulf, much of what transpires in the 1990s will be determined by the ability of Iraq and Saudi Arabia to coordinate their oil strategies. However, the summer of 1990 introduced the unprecedented aggressive tactics of Iraq in its effort to raise prices, its invasion of Kuwait, and its military threat to Saudi Arabia, which all make a unified price policy more difficult to achieve. Nevertheless, the future of OPEC will depend chiefly on the determination of these two giant suppliers acting in support of cartel policies. But as key members develop new interests downstream (and most cartel members lack the means to do so), unity will be ever more difficult to achieve. There is today no rallying leadership among its members. Without it, there is little chance of oil producers adopting the much-feared strategy of using their low-cost oil to drive non-OPEC suppliers out of business.

Circumstances affecting oil are always in flux, as the papers in this volume make clear. But the fundamentals of international energy trade—by whom prices will be set, who needs imports of what increasing volume and from which sources—must remain of continuing interest and even of mounting concern to the United States and most of its allies.

Were supplies to be obtained under truly market conditions, the prospect could be different. But they will not be. The underlying price of world oil will still be set by OPEC, which means by governments and their oil company agents. Perhaps the key question for the 1990s is the extent to which governments' intervention in oil will still play an ever larger role in supplying oil and setting terms in international trade. In the 1980s, there was a widespread belief that oil supply would become tight by the mid-1990s. This is no longer thought so certain. But if OPEC cannot work in unity, oil prices could be set increasingly by consumer markets.

There has never been a genuinely free market in oil, certainly not in the time of the oligopoly of the "Seven Sisters" (1950 to 1970) and certainly not since the early 1970s as OPEC member governments took control over the international crude oil trade. Despite recent investments downstream by key members of OPEC, not one of these countries has the slightest wish to subject its own most basic interest—increasing oil revenues—to the "market," but they may have no option.

Without a doubt, OPEC's search for unity depends increasingly on the relationships between Iraq, Iran, and Saudi Arabia and on the possibility that they cannot contrive to limit the amount of oil they put into international trade. Without that basic requirement met, the future of international oil trade might well become one of almost permanent glut and low prices. For the United States, it appears highly probable that its own oil production is started on a long-term decline. Moreover, for the foreseeable future, the nation's needs for ever more barrels will be met from the Middle East. Historically, this kind of situation has been the prime cause for U.S. government intervention to protect domestic producers from lower-cost imports and to deal with national security implications. We shall certainly be dealing with these two themes again. We will also have to face the repercussions in the Gulf from the presence of large U.S. armed forces interposed between Iraq and Saudi Arabia.

Nevertheless, the United States has never had a comprehensive, durable national energy policy and it could fail again unless several quite crucial forces come into play: (1) a president who is forcefully and permanently committed to protecting the energy security interests of the nation, (2) a Congress whose leaders in the Senate and in the House reinforce the president's commitment, and (3) both branches of government willing to establish energy priorities and programs intended to last beyond short-term political changes.

We have never been able to achieve these objectives. The country is itself divided with conflicting regional interests, particular energy interests clamoring for attention, and the public more or less unconvinced of the need to meet these objectives through long-term energy commitments. Some argue that only in circumstances of general war can the national will be mobilized. But the inescapable essence of "energy" is that years of large prior investments are required to improve upon the nation's long-term energy security. Others will assert with some justification that once the U.S. government becomes involved, too many interests become adversely affected. This need not, of course, be the result, assuming strong political leadership and expert direction of energy. U.S. energy policy is no task for amateurs.

The essays which follow deserve careful reading, for if government is called upon again to intervene in the interest of national security, the evolving circumstances of today's oil market, with observations on the recent past, will be crucial in deciding what we do. Industry will benefit from these essays, for they clarify a complex situation which has never stopped evolving.

It has been my good fortune for some years to have assisted Professor Wilfrid L. Kohl's International Energy Program at the Nitze School of Advanced International Studies, the Johns Hopkins University, as chairman of its advisory committee. Along with the others who have benefited from his seminars, I thank him for his able and objective direction of the only ongoing energy program in Washington, D.C., capital of the world's largest consumer of energy, the second largest producer of oil and gas, the largest importer of oil, and one of the largest importers of gas.

MELVIN A. CONANT

Preface

THIS book analyzes the behavior of the world oil market in the 1980s as a basis for understanding the world oil market of the 1990s. A major focus is the oil price collapse of 1985–86, when prices dropped from about $27 per barrel to below $10 per barrel within the space of a few months, then recovered to a range of $15 to $18 per barrel in 1987. The equivalent of a third oil shock, this precipitous price drop brought short-run benefits to energy consumers but raises serious problems for all oil-producing countries and potentially for U.S. and international energy security. To explain how it happened is the initial purpose of our study. We then move on to consider its implications for the future of OPEC, the United States, and the world oil market. A preliminary assessment is also offered of the impact of the Persian Gulf crisis of 1990–91 on the oil market, OPEC, and energy security.

This volume is a product of the International Energy Program (IEP) of the Johns Hopkins Foreign Policy Institute at the Paul H. Nitze School of Advanced International Studies in Washington, D.C. The IEP sponsors teaching and research on energy policy analysis and national and international energy issues. Increasingly, it is also expanding its focus to include the environmental impacts of various forms of energy use. The program organizes monthly professional seminars and periodic conferences and publishes a series of energy papers as well as occasional book-length studies. The IEP receives support from the U.S. Department of Education and from over a dozen corporate sponsors and corporate participants.

Most of the papers in this book were originally presented and discussed at international energy seminars attended by government officials, congressional staff representatives, industry experts, consultants, and academics during the period 1986 to 1989. We are grateful to the corporate sponsors and participants of the IEP for making this work possible without at any time interfering with its content and to participants in the seminars for their advice and suggestions. Melvin A. Conant, chairman of the IEP advisory committee, has for several years offered his leadership and counsel. We are indebted to him for his guidance of the program and for the foreword he has contributed to the book. For a number of years the Johns Hopkins Foreign Policy Institute, chaired by Harold Brown and directed by

Simon Serfaty, has provided a hospitable environment for all the activities of the IEP, including this book.

Several anonymous reviewers for the Johns Hopkins University Press provided helpful critique and suggestions as this book neared its final form. The keen interest in this project shown by Anders Richter, senior editor at the Johns Hopkins University Press when he retired at the end of 1989, was critical and is deeply appreciated. The steady guidance of Henry Y. K. Tom, who assumed editorial direction of the project, has also been vital in bringing the book to fruition. Finally, thanks are also due to Carol W. Rendall, IEP associate, who edited several of the papers, and to Audrey Abraham, Barbara Johnson, and A. Camille Richardson, who served as IEP program assistants since this project began.

Abbreviations

ANS	Alaska North Slope
ANWR	Arctic National Wildlife Refuge
CAFE	Corporate Average Fuel Economy
CPE	centrally planned economies
DOE	Department of Energy
DOJ	Department of Justice
EOR	enhanced oil recovery
EPC	Economic Policy Council
ERT	Unión Explosivos Río Tinto
FDIC	Federal Deposit Insurance Corporation
FRS	Federal Reporting Service
FTA	Free Trade Agreement
GCC	Gulf Cooperation Council
HHI	Herfindahl-Hershman index
IDC	intangible drilling cost
IEA	International Energy Agency
mb/d	millions of barrels per day
MOIP	Mandatory Oil Import Program
NCW	non-Communist world
NES	National Energy Strategy
NNP	net national product
NOC	National Oil Company
NRC	Nuclear Regulatory Commission
NSC	National Security Council
NYMEX	New York Mercantile Exchange
OAPEC	Organization of Arab Petroleum Exporting Countries
OCS	Outer Contintental Shelf
OECD	Organization for Economic Cooperation and Development
OIIC	Oil Investments International Company
PNOT	Philippines National Oil Company
PRT	Petroleum Revenue Tax
SPR	Strategic Petroleum Reserve
TPE	Total Primary Energy Consumption
UAE	United Arab Emirates
USGS-MMS	U.S. Geological Survey and Mineral Management Service
VLCC	very large crude carriers
WTI	West Texas Intermediate
WTS	West Texas Sour

WILFRID L. KOHL

Introduction

IN contrast to the tight market of the seventies, the international oil market of the eighties was characterized by falling demand for oil and downward pressure on oil prices due to a surplus of oil production. OPEC consistently underproduced in relation to its production capacity and lost market share to a burgeoning number of non-OPEC-producing countries such as Norway, the United Kingdom, Mexico, Egypt, and India. Indeed, OPEC has had great difficulty in managing this "buyers' market." In its efforts to do so, the cartel had to lower oil prices for the first time from $34 to $29 in March 1983 and establish production quotas for its members along with a collective output ceiling of 17.5 million barrels per day (mb/d). A year later, in 1984, the official selling price was decreased again to $28 before it tumbled out of control following the abandonment of fixed prices in 1985–86.

In Chapter 1, Fadhil J. Al-Chalabi, OPEC's deputy secretary general from 1978 to 1989, explains OPEC's role in the oil price collapse. With unusual candor, Chalabi states that OPEC's pricing policies of the 1970s and early 1980s "were self-defeating for its members. . . . Very high prices brought OPEC oil under an increasing pressure, which led to a continual fall in the organization's share in the world market." A series of structural changes followed in the world energy economy which worked against OPEC's interests—conservation in industrialized countries, fuel switching, and a substantial rise in non-OPEC production. Chalabi reviews the history of OPEC price setting in the world oil market, pointing out that one major cost has been that OPEC members were deprived of the ability to compete with the new non-OPEC producers because of their obligation to defend official prices. Thus, OPEC became the "residual supplier" of world oil.

In mid-1985 Saudi Arabia, which had previously performed the role of "swing producer" in OPEC until its production share declined to about 2 mb/d, announced a major shift in its oil policy: the adoption of "netback pricing." In December, OPEC decided to abandon fixed prices to expand its market share. These actions unleashed the oil price collapse. But, as Chalabi points out, OPEC was neither clear about nor prepared to implement this strategy. The result was chaos and the precipitous drop in oil prices.

With their revenues from oil sales sharply diminished, the low reserve–high population countries were the real losers in this strategy, which unleashed major new tensions within OPEC. It also found opposition among the industrial countries and from other oil-producing developing countries. In the end OPEC could not stand up to these pressures and was forced to change course. By December 1986 OPEC had returned to fixed prices and production quotas.

But market fundamentals did not change. The fixed price system did not last long, as it was impossible for OPEC to defend both volume and price at the same time. Chalabi goes on to explain how OPEC, assisted by growing demand for its oil, shifted to a market-oriented pricing system in 1987–88. With a return to higher levels of production, prospects for OPEC at the end of the 1980s and the beginning of the 1990s looked much brighter. Critical factors in OPEC's future are how fast world demand for OPEC oil will increase and how much OPEC countries themselves will be willing to invest in new production capacity.

The next several chapters elucidate the framework set out by Dr. Chalabi and amplify particular factors which contributed to the oil price collapse, thereby placing it in broader perspective. Saudi Arabia, the country with the largest oil reserves in the world, plays a critical role as OPEC's most important member. Hossein Askari, a professor of international finance who has lived and worked in Saudi Arabia, reviews the history of Saudi oil policy and discusses the economic and political factors which have shaped it. He focuses especially on the importance of oil revenues and oil investment income to the Saudi economy. In 1985 Saudi Arabia, which had already begun to draw down its reserves, was hurting financially and could not meet its short-run revenue requirements, especially subsidies to the economy and military and foreign assistance expenditures. Hence, the dramatic shift in oil policy and the abandonment of its role as swing producer.

Askari contends that Saudi oil policy in the past has worked against the country's economic interests, as the Saudis have frequently imposed production restraints on themselves in order to defend OPEC prices that were too high. He thinks the Saudis have learned a lesson, but have they? As Askari points out, the political necessity within OPEC of accommodating expanded oil production by Iran and Iraq after their eight-year war promises to provide a severe test of Saudi policy in the future and perhaps edge it back again toward a swing producer role in the cartel.

Oil consultant Edward N. Krapels provides a comprehensive view of the market fundamentals—supply, demand, and the role of oil inventories—that shaped the world oil market in the 1980s. He argues that four fundamental factors explain the fall of oil prices during the decade: (1) the increase in surplus oil production capacity in the world coupled with a decline in oil demand below a level that important OPEC members could not tolerate; (2) the decline in financial strength of OPEC and the oil industry; (3) the demise of vertical integration in the oil industry, which meant that henceforth oil was traded in open markets in spot sales or under short-term contracts; and (4) the absence of an institutional mechanism to bring supply and demand into balance. He concludes that the interplay of these fundamental factors portends continued oil price instability in the 1990s.

A historical perspective on the history of oil prices is presented in Chapter 4 by David H. Vance, an energy economist in the research division of the U.S. Department of State. He demonstrates that the high oil prices that emerged following the two oil shocks in the late seventies and early eighties were an aberration. An eventual price collapse was to be expected. Vance reviews the world oil production profile developed by geologist King Hubbert, as modified in a more recent projection by Leonidas P. Drollas, to highlight the large potential excess capacity in the world oil market that can be expected to exert downward pressure on oil prices over the next several decades. Vance also argues that market forces rather than price controls are more likely to achieve price stability in the oil market.

OPEC, with its dominant position in world-proven reserves, will try to hold prices above the long-run historical level. But if OPEC discipline falls apart, prices could again plunge as they did in 1986. According to Vance, the basic security problem for the United States is that of increased import dependence on potentially insecure Middle East oil sources, but the efficient utilization of world oil resources leads inexorably to such dependency. As insurance against future market disruptions, he recommends a policy of strategic oil stocks and reliance on the International Energy Agency, not an oil import fee.

In a related analysis, energy economist and consultant Philip K. Verleger explains important structural changes in the world petroleum market in the 1980s. He first reviews the declining role of the major international oil companies in the production and refinement of crude oil and the breakdown of integration. A related development is the decline in concentration in the oil industry and the increase in competition, which, he argues, created the conditions that precipitated the 1985–86 collapse in oil prices and the ensuing period of greater price volatility. According to Verleger, the degree of concentration in the industry (i.e., market share) is more important than surplus capacity of producers in determining future oil prices.

Verleger then describes the growth of commodity market institutions in the oil sector, which is linked to the decline in integration in the industry. He explains the important role of spot markets, forward markets, and futures markets in setting oil prices now and in the future. These institutions grew up to reduce the exposure of buyers of crude oil to the risks of price fluctuations after OPEC was no longer able to manage the world oil market. Today oil has become the largest of all spot and forward markets.

The next three chapters focus on the impact of the oil price collapse on U.S. oil production, the U.S. oil industry, and U.S. policies on energy security. Economist Carol A. Dahl examines U.S. oil production in the eighties within the context of historical data on production and reserves and compares U.S. and world oil production. She finds that high oil prices at the start of the decade gave rise to a temporary drilling boom that led to a surge in oil production and reserves. However, as oil prices declined, so did U.S. production. By the end of the decade U.S. shares of world production and reserves had resumed the downward trend begun a decade before. Her analysis of U.S. and international cost and investment patterns suggests that the role of the United States in the world oil market will continue to decline.

The price collapse caused a major contraction in the U.S. oil industry, as examined by Dillard P. Spriggs, an experienced financial analyst. Many independent companies were forced out of business. The oil service sector was particularly hard hit, and the number of drilling rigs plummeted. Major integrated companies, however, fared better because their substantial downstream (refining and marketing) operations profited from lower raw material costs. However, even the major oil companies had to curtail drastically their exploration and production budgets, and all companies have had to learn to operate more efficiently.

The price collapse gave further impetus to a wave of mergers in the oil industry. Echoing one of the conclusions of the Dahl chapter, Spriggs also finds that the high costs of U.S. production are motivating many companies to concentrate their future exploration and production expenditures abroad, reinforcing a likely trend toward rising U.S. oil imports.

Chapter 8, written by a political economist, explores the recent U.S. policy debate on energy security provoked by the oil price collapse. The *Energy Security* report by the Department of Energy and other studies have expressed concern over the depressed state of the U.S. oil industry and the prospect of sharply rising oil imports in the years ahead. But the Reagan administration took no major new policy actions on either front after oil prices recovered in 1987 to the $16 to $18 range. The administration strongly opposed an oil import fee, and Congress so far has not favored such an action. However, the Bush administration may reconsider policies to promote U.S. energy security.

Following an analysis of the economic, military, and foreign policy dimensions linking oil and U.S. national security, this chapter considers in some detail available policy options. It concludes that U.S. oil imports are only one among several factors that condition U.S. and international energy security, which depends more fundamentally on the stability/instability of the world oil market and oil prices at any given time.

In Chapter 9, Edward L. Morse and Julia Nanay, drawing on their special knowledge of current industry trends as international oil consultants, describe how the oil price collapse accelerated major trends that had been developing in the oil industry toward increased competition. They focus on the responses of the oil exporters, both within and outside of OPEC. First, they explain the new producer country pricing policies designed to maintain, recapture, or increase market share. After a critical discussion of administered pricing, which the authors believe has severe disadvantages and will not likely be followed by many OPEC members, other pricing methods currently in use are analyzed. These include auctions and spot sales, prices linked to specific types of crude oil, market-basket pricing, netback pricing, retrospective pricing, and processing and tolling arrangements.

The second kind of response is downstream market reintegration. Some of the wealthier OPEC exporters are establishing direct ties to refiners in an effort to gain more assured access to markets. The actions of several countries are reviewed, notably Kuwait, Venezuela, Libya, Nigeria, Mexico, and most recently Saudi Arabia, which in 1988 purchased a large joint venture stake in Texaco's refining and marketing system in the United States. The authors conclude that while the new pricing policies being pursued by the producers seem to ensure more price volatility in the oil market, market reintegration could work in the other direction, toward price stability. The outcome of the latter is not yet clear, but reintegration is certainly yielding a radical restructuring of relations between national oil companies in the producer countries and private companies in the consumer markets.

In the penultimate chapter, written before the Persian Gulf crisis, internationally acclaimed oil industry analyst John H. Lichtblau takes a look at the future of the world oil market. He begins by posing the question, does the 1986 price collapse signal the end of the OPEC cartel? Referring back to Dr. Chalabi's opening chapter, Lichtblau agrees that the price collapse was mainly caused by the actions of OPEC; that is, it was self-inflicted. He then discusses how the oil market has responded, noting that world demand for oil has reversed and actually increased in the years 1986 to 1988. Turning to his forecast for the future, Lichtblau sees oil prices determined by the interaction of the OPEC cartel (perhaps supported by some non-OPEC exporters) and market forces. In time the strength of OPEC will be gradually reinforced by market fundamentals as world demand increases slowly and prices move gradually upward. The end of the Iran-Iraq war complicated OPEC's problem of trying to manage the market, since substantial additional supplies began to be made available by these two former combatants, price volatility continues to be likely as substantial excess producing capacity will remain available while OPEC and the market battle it out.

Iraq's annexation of Kuwait in August 1990 and the subsequent U.N. trade embargo that cut off Iraqi and Kuwaiti oil exports to the world market ushered in the first oil crisis of the 1990s. Iraq's action, the subsequent massive deployment of U.S. and allied troops on the Saudi border, and concern regarding the outbreak of hostilities that might damage Saudi oil fields and refineries, led to a doubling of oil prices. The war that began in January 1991 dramatizes more clearly than ever before the dangers of dependence on Persian Gulf oil which continues to be vulnerable to unpredictable political events. The background of the latest Gulf crisis, the resultant oil price shock, and the implications for OPEC and the future of the oil market are the subject of an Epilogue, Chapter 11.

After the Oil Price Collapse

FADHIL J. AL-CHALABI

1. The World Oil Price Collapse of 1986: Causes and Implications for the Future of OPEC

IN the summer of 1986, the world was stunned by the sudden oil market turbulence, in which Gulf prices plummeted at breathtaking speed to a range of about $6 to $9 per barrel after having been sustained at around $28 per barrel for many years, up to December 1985. The price collapse and market chaos were short-lived, however; the falling trend in oil prices was reversed in September 1986, when an OPEC "interim agreement" of production restraint was concluded. Prices were stabilized in the beginning of 1987 somewhere in the middle of the above two points, that is, at an average of about $18 per barrel (for a basket of six OPEC crudes and one non-OPEC crude, the Mexican Isthmus),[1] but only temporarily. The conclusion of OPEC's December 1986 Agreement, which finally ended the "price war," or the OPEC market share strategy, and reinstated the fixed price system, which had been abandoned de facto in 1986, was achieved after a series of negotiations at the highest political levels within the organization and after contacts with many non-OPEC countries had been made.

These tumultuous price movements clearly showed how crucial OPEC's decisions are, not only in shaping world energy markets but also in affecting, in a far-reaching manner, related sectors, especially world financial indebtedness; many highly indebted countries from the Third World, such as Mexico, Egypt, and Venezuela, are oil-producing/exporting countries.

After twelve years of oil price administration by OPEC, based on an uneconomic and rigid fixed pricing system, the organization suddenly decided in December 1985 on a de facto abandonment of that system in search of a higher market share. This happened after OPEC countries, especially those with high reserves, had seen their share in the world oil market dwindle dramatically for several years before, mainly as a result of that price system as well as the overpricing of oil by OPEC. What restored partial stability (which later proved to be precarious) to the world market in 1987 was simply the fact that OPEC returned to the old system of fixed prices by formally abandoning the short-lived market share strategy, but at lower price levels than prior to 1986.

However, it took less than a year before the re-adopted fixed price system started to crack under market realities, finally causing its total collapse. This led ultimately to OPEC's applying, in January 1989, a new system whereby prices are no longer fixed and adhered to by member countries. The new price of $18 per barrel became a notional or "target" price that OPEC strives to reach successfully or unsuccessfully through production regulations. With the new system the price defense mechanism of OPEC shifted finally from the unrealistic defense of an abstract and theoretical price, fixed by the organization irrespective of market realities, into market-oriented pricing systems formulas.

OPEC: "Price Maker" or Defender of Its Market Share?

The history of oil price making since OPEC took over pricing from the international oil companies in 1973 shows clearly that stability or turbulence in the world market would depend to a very great extent on the organization's decisions concerning price making and/or the production of oil. More important is OPEC's approach to a basic economic concept which is related in the final analysis to income maximization: whether OPEC behaves in the world oil market as a "price maker" (in which case income maximization would depend more on the defense mechanism of the price of oil — the maximization of its per barrel income) or a "price taker," accepting the price determined by the market and resulting from the interplay of market forces (the maximization of income through volume). According to this approach, income maximization could be achieved through a higher market share. In the former case, the organization sets the price to be taken by other operators, notwithstanding the effects of the price on the call on its oil or its market share, whereas in the latter case, OPEC would defend greater export volume (market share), leaving the price to be determined by the market forces without any defense mechanism by OPEC. Rigidly taking either of these two options would lead, in a situation of oversupply, to far-reaching effects on the income maximization of OPEC countries. Defending the price would mean allowing the volume to decline substantially in response to market forces, whereas defending the organization's market share would mean leaving the price to deteriorate in response to the oversupply.

The defense price mechanism adopted by OPEC since the 1970s was to set a "take it or leave it" price by which its members should abide (theoretically) regardless of the prevailing market prices, which in most cases are persistently below OPEC official prices. Conversely, OPEC could adopt a competitive approach vis-à-vis other suppliers of energy in a quest to move larger volumes of its oil at market-oriented prices. In the first case, OPEC secures price stability for the world oil market at the cost of losing market share to the other producers, whereas in the second, OPEC can secure a market share for itself but with the risk of provoking a price collapse.

The answer to this question is crucial, especially when it comes to the second option; that is, OPEC adopting a purely competitive approach vis-à-vis other producers of energy, including non-OPEC oil. Here, the adverse effects on market stability are tremendous if we take into account the fact that OPEC's oil, especially in the Middle East, is the cheapest in the world to produce and that low OPEC production levels such as the one in 1985, when OPEC's production declined to 15.5 mb/d, represent about half the organization's sustainable production capacity. Such an approach would simply mean that prices in the world market could fall to less than $4 to $5 per barrel, being the average world operation cost of producing a barrel (assuming that the capital cost is either recovered or sunk). Conversely, taking the other approach of rigid price defense at the expense of OPEC's share of the world market would simply mean that member countries' production could dramatically fall, as indeed it did when OPEC's production fell in 1985 to half its production peak of 31 mb/d in 1979. In this case, OPEC's capacity to administer the price or influence its making would be enormously reduced.

OPEC's Earlier Price Setting Leads to a Loss of Market Share

As already mentioned, the OPEC pricing mechanism until the end of 1985 was to set a floor price for reference crude oil (until then Arab Light 34′ API ex Ras Tanura, called the OPEC marker crude) under which no member country was supposed to sell its oil. This price was called the official selling price on the basis of which prices of other crudes were fixed, taking into account the relative values of its various crudes based on the difference in quality and geographic location. This system, which was applied for more than twelve years, proved to be uneconomical and was detrimental to the organization's basic interests, especially for those countries in the Gulf which account for more than 60 percent of the world's proven reserves.

As a result of the fixed price system at very high levels adopted by OPEC during the 1970s and the first half of the 1980s, world demand for oil fell dramatically, whereas supplies of oil outside OPEC started to grow at an accelerated pace. The result was that the call on OPEC oil fell drastically in terms of export volumes, simply because in a situation of falling demand, non-OPEC producers, while benefiting from high prices secured to them by OPEC, were able to replace OPEC oil in the market as much as they could by simply discounting the OPEC price. Whereas most of the members of OPEC abided rigidly by a fixed price, the non-OPEC producers enjoyed total freedom in adjusting their contract prices in a way that secured for them a complete market competitive advantage.

It is true that this system secured price leadership for OPEC in setting the price level for the market. But the self-imposed price discipline deprived OPEC members of the flexibility to compete with other oil-producing/exporting countries. They had to abide by the floor price while non-OPEC energy suppliers were selling at competitive market prices, which, in a glutted market, were generally lower than OPEC's floor price. In such a situation, the OPEC pricing mechanism meant that the organization would have to act in the world energy balance as the swing producer or the residual supplier of energy in order to achieve an equilibrium between world supply and demand at the price level it sets. Buyers would first take all their energy supplies, including oil, from outside OPEC before turning to the organization's member countries to buy volumes that would meet the difference between the world energy requirements and the non-OPEC supplies. Naturally, in a glutted market this system represented unfair competition between OPEC member countries and the other energy suppliers and would inevitably lead to a continual decline in OPEC's share of the world energy market, since all non-OPEC suppliers can maximize their market share volume at the expense of OPEC.

Fear of Depletion Leads to Noneconomic Pricing

OPEC's pricing policies of the 1970s and early 1980s, which enabled its average price to reach the staggering level of $36 per barrel (the upper band of the then existing two-tier price system),[2] were self-defeating for its members, especially for those with enormous proven reserves and increasing export capacities. Very high prices brought OPEC oil under an increasing pressure, leading to a continual fall in the organization's share in the world market. Although subsequently the increase in OPEC production from to 15.5 mb/d in 1985 to 20 mb/d in 1988 helped a great deal in attenuating OPEC's problems of quota distribution, it did not solve them completely. The adverse volume effects of the earlier pricing practices are still lingering on

OPEC's capacity to administer the price of oil or influence its formation in this market, as will be discussed later.

What was especially important to OPEC's pricing policies during those twelve years seems to have been the organization's drive toward maximizing the per barrel income of its exports rather than considering long-term global oil income by taking into account the effect of volume as well as price. Equally important was the fact that this policy was initiated at a time when demand for OPEC oil was increasing to levels considered by its policy makers as very high, giving rise to fears of early depletion of member countries' recoverable reserves.

During the 1970s, there was a genuine fear in some OPEC circles of the impact of high production rates on the life-span of their reserves when in 1977 the reserves-to-output ratio of OPEC oil fell to its lowest level of about forty years, based on that year's production rates of and the then known proven reserves. In fact, OPEC policy makers developed an alarmist attitude vis-à-vis the growing world demand for its oil, an alarm that was fueled by false energy forecasts that flooded Western oil literature during the late 1970s and indicated an imminent shortage of oil. Indeed, many forecasts made in 1977–78 estimated world demand for OPEC oil as exceeding 40 mb/d by 1990, a magnitude of production which went far beyond the organization's capacity at the time. Naturally, should effective world demand for OPEC oil reach such magnitude, a real physical shortage of oil would occur, with soaring prices that could reach spectacular levels. Higher prices therefore were considered by OPEC policy makers as a means of not only increasing revenue but also of reducing demand for their oil in order to avoid the over-depletion of their reserves. This conservationist approach was used by the OPEC radicals as a justification for high price policies which irrationally followed market flare-ups, especially during what was called the "second oil shock," when the official selling price for the marker crude rose from $14.34 per barrel in January 1979 to as high as $36 per barrel in January 1981.

The reality, however, was that the noneconomic approach to OPEC's pricing policy was influenced more by the successive price increases in the spot market caused by political events, notably the Iranian Revolution in early 1979 and the Iran-Iraq war in 1980, than by any economic concept of price optimality under the conservationist approach. No less irrational price policy was that while OPEC's official selling prices were increased in light of the rising trend in the spot market, no subsequent action was taken by the organization to reduce the price in light of the downward trend in spot prices that followed in 1981 and thereafter.

High Oil Prices Induce Fundamental Changes in World Energy Supply and Demand

In this sense, therefore, OPEC was its own worst enemy by setting prices at high levels that triggered far-reaching structural changes in the world energy industry at the organization's expense. Three distinct developments occurred: a substantial decline in the rate of growth of energy consumption in industrialized countries, a change in the global energy structure toward lower share of oil, and a change in the world oil structure toward an increasingly lower share for OPEC.

The first development was the result of a decoupling between economic growth and energy requirements, mainly because of energy conservation efforts and a continual improvement in the efficiency of energy utilization, so that economic growth in industrialized countries required less energy consumption than hitherto. Conservation was suddenly considered to be the cheapest form of energy to compete with other commercial forms of energy, let alone expensive oil. This was enormously enhanced and accelerated by government policies in the OECD countries, where various internal fiscal and nonfiscal measures were taken that not only motivated consumers to save energy but also encouraged investment aimed at continually improving the efficiency of fuel utilization.

The second development was caused by the substitution of coal and nuclear energy for oil, especially in electricity generation and fuel burning for industry. High oil prices, coupled with government policy actions in industrialized countries, made those sources so attractive that a continual shift toward them occurred at the expense of oil. The oil product which suffered most from this process was fuel oil, and its consumption in OECD countries declined dramatically from about 11 mb/d in 1973 to about 4.75 mb/d in 1985. This happened as the consumption of coal and nuclear energy in these countries rose from 725.2 million tons of oil equivalent (mtoe) to 1,126.1 mtoe during the same period or an increase of 8.02 million barrels of oil equivalent per day (mboe/d). This substitution process led to a shrinkage of oil's share in the world energy mix from 55.28 percent in 1973 to 44.56 percent in 1985.[3]

The most important development in the world energy situation was the impetus which high prices gave to investment in exploration and production of new oil reserves outside OPEC, leading to the addition of about 8 mb/d of new oil which replaced OPEC oil. As a result, OPEC's share of total oil supplies declined from a peak of 65 percent (although many other estimates put OPEC's share at 68 percent) in 1973 to a low of 40.1 percent in 1985 (Table 1.1).

Table 1.1. World Primary Energy Production (excluding Centrally Planned Economies) (in millions of tons of oil equivalent (MTOE) and % shares)

	1973		*1979*		*1986*	
	MTOE	*%*	*MTOE*	*%*	*MTOE*	*%*
World Oil	2,432.4	56.4	2,626.6	54.5	2,131.1	45.6
OPEC oil	1,582.2		1,612.2		860.7	
OPEC %	65.0		61.4		40.1	
Solids	756.2	17.5	884.5	18.4	1,047.3	22.4
Gas	816.4	18.9	831.1	17.3	807.0	17.3
Nuclear*	41.4	1.0	127.4	2.6	284.2	6.1
Hydro*	262.9	6.1	348.3	7.2	407.9	8.7
Total†	4,309.3	100.0	4,817.9	100.0	4,677.5	100.0

Source: U.N. Energy Statistics Yearbook 1985.

* Reflects the primary energy equivalent of inputs of nuclear and hydro to electricity production.

† Total may not add up because of rounding.

The increasing pressure on OPEC was predominantly the result of the economic impact of the "second oil shock," which led to a sudden increase in the cost of world oil imports from $194.5 billion in 1978 to $413.4 billion in 1980. OPEC's income from oil rose from about $136 billion to the staggering figure of about $287 billion during the same period. This aggravated the adverse economic impact of the "first oil shock," when OPEC's oil revenues rose from about $24 billion in 1972 to about $120 billion in 1974. The economic effects of the sharp rise in prices and other political considerations were behind the political action taken by the industrialized countries (symbolized in the creation of the International Energy Agency [IEA]), whereby measures were adopted to restrict consumption and imports of OPEC oil in those countries.

Reducing Exports to Defend Prices Threatens Present and Future Prospects of High-reserve Producers

These developments had a twofold impact on OPEC's role as the main price arbiter in the world market. The first was of a long-term nature, adversely affecting not necessarily all OPEC's member countries but rather the more influential ones—the holders of large oil reserves, especially the Gulf countries. The sharp decline in OPEC's market share posed a real threat to the position of these countries in the world's energy arena, since they were more concerned with maintaining long-term demand for their oil, which was practically their only source of revenue. This was in sharp contrast with low-reserve member countries, which were more preoccupied with short-term financial requirements that could be more easily satisfied by higher prices.

The aforementioned structural developments in the energy situation led to a far-reaching negative impact on the organization's ability to administer oil prices and on the defense of price stability itself. In the 1970s, it was sufficient for OPEC to fix the price and defend it as a minimum floor level, leaving the volume to be determined by market forces. Because of the very long lead times for investments in energy, especially oil, which can be as long as seven to ten years before the new capacities come on stream, the short-term impact of higher prices on the volume and demand for oil was negligible. It was, therefore, very easy during the few postshock years for OPEC to defend the price at volumes which were still high enough to generate cash flow far in excess of its member countries' financial requirements, but only temporarily.

The problem arose when the price impact started to be felt and demand for OPEC oil began to decline sharply in 1981–82. Price defense meant, in this case, that OPEC had to continually sacrifice export volumes as the only means of maintaining price stability. Between 1979 and 1982, OPEC's total production fell from 31 mb/d to 19 mb/d, and the continual reduction in demand for its oil led to the organization adopting production control measures in order to defend the price. As of 1982–83, OPEC decided on an overall ceiling to its production, distributed into national quotas, beyond which no member country was expected to produce.

OPEC's Production Cuts Cause Internal Problems

As the deterioration in the market continued, the price defense policy required OPEC to reduce its production ceiling successively from 18 mb/d fixed in March 1982 to 17.5 mb/d in March 1983 and then to 16 mb/d in October 1984. This led to increasing tensions among member countries in reallocating quotas, as growing difficulty in maintaining production within limited quotas caused some member countries either to reject formally and publicly their quotas and therefore not abide by them or to accept the allocated quotas formally but to exceed them in reality while asking at the same time for higher ones. The United Arab Emirates (UAE), although having committed itself to OPEC's decisions, nevertheless felt that its quota was too low to be observed. Some smaller countries, notably Ecuador and Qatar, also expressed dissatisfaction with their quotas. Generally speaking, increasing financial hardship in most OPEC member countries pushed many of them to produce in excess of their quotas. OPEC's quota problems were further complicated by the Gulf war, especially as far as Iraq's quota is concerned.[4]

Saudi Arabia: Problems Mount for the Swing Producer

In spite of these problems, the price of $28 per barrel held until December 1985, mainly because in accepting the role of swing producer within OPEC the major producer, Saudi Arabia, was absorbing, in accordance with the London agreement of March 1983, the decline in world demand for OPEC oil through a reduction in its own production. When OPEC agreed on an overall ceiling of 17.5 mb/d in the aforementioned agreement, no specific quota was allocated to Saudi Arabia, contrary to all other member countries, whose total national quotas amounted to 12.5 mb/d. The 5 mb/d difference between the overall OPEC production ceiling of 17.5 mb/d and the total of those of twelve countries' production (12.5 mb/d) was considered to be a "production swing" to be allocated to Saudi Arabia to vary up and down in the light of market developments and within the overall ceiling of OPEC production. This meant that, when demand for OPEC oil fell short of the global ceiling of 17.5 mb/d, the twelve OPEC members would produce their quotas totaling 12.5 mb/d, whereas Saudi Arabia, in accordance with the 1983 agreement, consented to swing down its production to absorb the difference. The fall in OPEC production below the ceiling meant that it was mainly the Saudi production which had to swing down to absorb the difference and not all the organization's members. This system put Saudi Arabia under great pressure as the main price defender within OPEC. While other countries could produce or even exceed their quotas, Saudi Arabia had to absorb practically the bulk of the downward swing in the total OPEC production. In practical terms, the continual decrease in world demand for OPEC oil meant a continual decline in Saudi Arabia's production until, in the summer of 1985, it reached a level marginally above 2 mb/d, or about half the amount of the production swing (Saudi quota) of 5 mb/d; in fact, Saudi Arabia had to swing down its production in the defense of the price because of the structure of its exports as well as its willingness to play the role of a swing producer under the terms of the 1983 agreement. This is simply because Saudi Arabia is the most important crude exporter which had to abide by the official price for its crude. Among the causes of that disproportionate shouldering of the burden of price defense among OPEC member countries was Saudi Arabia's pattern of crude marketing compared to that of other members.

The bulk of Saudi oil exports takes the form of crude oil lifted at official selling price by the former shareholders of Aramco. Furthermore, more than 60 percent of its crude exports are of Arabian Light, which, until December 1985, served as the official marker crude for all OPEC. This meant that Saudi Arabia's crude exports always had to be lifted at the official selling price, which in a glutted market is generally higher than the market's ongoing price, thus providing fewer incentives for buyers to lift from that country. By contrast, the pattern of marketing oil in many other member countries provides much greater flexibility for lifting. In countries like Algeria, Kuwait, and Venezuela, the bulk of oil exports is exportable in the form of refined products, which are not subject to the OPEC pricing regime and therefore are priced unilaterally by the producing countries in light of the prevailing market price. Such marketing patterns give those OPEC countries a significant marketing flexibility and competitive advantage over other countries which sell crude only at official prices. Furthermore, some countries such as Nigeria, the UAE, and Libya still have part of their oil lifted under concessionary terms, which generally secure for the concessionaire a certain profit margin. In times of oversupply, the profit margin can be adjusted bilaterally between the host government and the concessionaire in such a way as to provide incentives for the latter to lift more oil without the adjustment being controlled by OPEC. This obviously meant that these countries can also enjoy a certain competitive advantage in selling oil compared with others which no longer have concession oil, because of either the nationalization of their oil or the government takeover of the oil industry by agreement with the multinational companies in the mid-1970s.

Other forms of marketing flexibility for some OPEC oil also exist such as Venezuela's extra heavy crude not being subject to OPEC's price regime. In fact, this advantage, besides the aforementioned one concerning the export of refined products, puts this country at a special competitive advantage compared with any other OPEC country, since it is estimated that as much as 90 percent of Venezuela's exports could be moved in the world market at competitive market-oriented prices.

Against this disparity of marketing patterns in member countries, many countries introduced a number of devices aimed at increasing the competitiveness of their oil. These included barter deals, countertrade, and processing deals.[5] Such deals made oil exported from OPEC zones more competitive in the market than others, and thus the countries resorted to these competitive devices to fully produce and sell their quotas. Saudi Arabia and a few other OPEC members whose petroleum exports are mainly in the form of crude oil were therefore left with such a competitive disadvantage in marketing their crude that they had to choose between two options: Either they had to adhere to the official selling price and see their export volume dwindle from then on, in which case they help and in fact support world market price; or they had to concede to buyers marketing advantages similar to those offered by the other OPEC countries in the form of price discounts, nominal or real, outright or concealed through many devices that would provide enough incentive to lift oil. But in this case prices on the world market would erode.

Saudi Arabia's rigid marketing system and the fact that its Arabian Light oil served as the marker crude for pricing OPEC's oil, in addition to its acceptance of the role of swing producer within the organization, had made it the only important OPEC member to abide by the official prices. This situation forced Saudi Arabia to be the principal absorber of the OPEC "volume shock." The financial pressure that arose inside the country as a result of the decrease in its production prompted the Saudis to formally announce their rejection of the swing producer role within OPEC, however, and their adoption of a position in which they could produce their quota in the same manner as other member countries were doing. (In an overall production ceiling of 16 mb/d until the end of 1985, the Saudi "quota" was about 4.353 mb/d.)

By the middle of 1985, it became obvious that with the continual fall in world demand for OPEC oil and the new policy of Saudi Arabia the price mechanism could no longer be sustained. The trigger to the breakdown of the old system was Saudi Arabia's decision to no longer adhere to the official selling price and to adopt market-oriented pricing formulas which would enable it to sell its entire quota. The new pricing system, based on netback values, was nothing more than an attempt by that country to put itself more or less on a par with many other OPEC countries as far as flexibility in marketing oil was concerned.[6]

The Impact of Increased Non-OPEC Production

From the above analysis, it seems clear that the real cause of the OPEC price shock was the continual fall in the organization's production that started in 1980. OPEC's major concern was not so much the dramatic fall in world oil consumption as a more important and immediate factor: the continual rise in oil production and exports outside OPEC. Exorbitantly high prices made investment in high-cost oil-producing areas profitable, so the enormous capital investment made in the North Sea, Asia, and Africa was recovered in a very short time. Higher prices also guaranteed more than adequate profit margins for investment in the United States in expensive and sophisticated enhanced recovery methods and in continuous drilling to maintain high levels of production from an extremely limited resource base. This served to counteract the natural falling trend of U.S. production, which, without higher prices and hence investment, would have declined noticeably. Table 1.2 shows how OPEC and non-OPEC production and exports have developed since the so-called first oil shock.

OPEC's real problems started in 1982 when non-OPEC production had increased to such a level that it exceeded the organization's production. Among the reasons that explain how non-OPEC oil was replacing that of OPEC is the former producers' aggressive marketing policies. While OPEC abided by a self-imposed production restraint program and sold a great part of its crude oil at official selling prices, non-OPEC producers accepted no limitations or restrictions on their production levels sold at competitive market prices. In fact, non-OPEC oil-producing/exporting countries had been systematically undercutting OPEC's price in order to maximize their market share at the expense of the organization. This not only had the direct effect of displacing OPEC oil in the world market but it also created within

Table 1.2. OPEC and Non-OPEC Oil Balances, 1973–1986 (in millions of barrels per day)

	Crude Oil Production		*Net Oil Exports of Net Exporting Countries*	
	OPEC	*Non-OPEC*	*OPEC*	*Non-OPEC**
1973	30.99	14.90	29.42	1.40
1974	30.73	14.46	29.01	1.21
1975	27.16	14.26	25.47	1.24
1976	30.74	14.40	29.12	1.47
1977	31.25	15.36	29.14	1.64
1978	29.81	16.57	27.74	1.90
1979	30.93	17.70	28.66	2.14
1980	26.88	18.36	24.46	2.65
1981	22.60	18.84	20.00	3.15
1982	18.99	20.09	16.03	3.98
1983	16.99	20.88	14.05	4.84
1984	16.35	22.20	13.87	5.14
1985	15.45	23.05	12.88	5.66
1986	18.33	22.63	15.45	5.20

* Excluding centrally planned economies.

OPEC the feeling that the organization had been sacrificing volume share in the defense of a price, which benefited other oil producers at no cost to them. This situation prompted many OPEC policy makers to question the wisdom of the organization's entire pricing system, which had led to such a great loss of its world market share.

The oil situation, which was polarized prior to OPEC's reversed "price shock" of 1986, can be summarized in the following terms. By accepting the role of world energy swing producer, the organization had in effect to support, single-handedly, price structure for the world oil market which caused oil and nonoil energy investments outside OPEC to flourish. It was the decrease in OPEC production, in support of the world oil price structure, that resulted in an increase in oil and nonoil energy production outside OPEC.

Obviously, the new situation represented an increasingly inequitable distribution of the gains from world energy trade between OPEC and non-OPEC energy producers. Weakened by the increased financial problems of its members and by the mounting political dissensions and conflicts among them, the burden of OPEC's production decline became too heavy to be borne by OPEC alone. Furthermore, within the organization itself the growing burden of price support was not evenly distributed among its members, since Saudi Arabia was "swinging" down proportionately more than the others. Never in its history had OPEC looked so weak and fragile as in the summer of 1985, when its inner tensions peaked. The large producers, especially Saudi Arabia, felt that they were bearing a greater portion of the price load than others and that they had reached a point beyond which they could not continue sacrificing, perhaps undoing their national interest, while the small producers were apprehensive that Saudi Arabia could drop its support for the price.

Increasing Revenue Loss and Rising Internal Tension Result in a Shift to a Market Share Strategy

With the continued decline in the organization's production to the advantage of non-OPEC producers, it became apparent in December 1985 that OPEC's price administration at that high level was no longer possible. A change of policy was deemed necessary to reverse the trend. OPEC had to find the means to solve what seemed to be its major problem: to increase total production to a level high enough to meet the member countries' minimum quota requirements. The market share strategy was announced and implemented but for only about eight months of 1986, from January to August. OPEC was neither clear about nor prepared for the specific objective of this strategy, nor was it united on a well-defined program to implement the strategy. When the organization declared its intention to obtain a fair market share, no definition was given of what OPEC considered a fair share, except for the vague notion that "it should be consistent with its member countries' financial requirements for development." When OPEC tried to define its objectives, no agreement was reached within the organization.

Although officially no resolution was passed to abandon the fixed price system, OPEC's actual behavior during these eight months was totally market-oriented, which meant a de facto abandonment of the fixed price. OPEC countries were all selling under various price formulas that would guarantee for them the sale of their production. As if the shift in price making was not enough to secure competitiveness for OPEC producers, they started gradually to ignore any restraints on their production also. For some time, the OPEC producers were behaving like those of non-OPEC countries. Given the disparity in cost and capacity between the two categories, the downward price spiral had to get in action at an accelerated pace. As a consequence of this free-for-all situation, the price of Arabian Light had to fall to a range of $6 to $9 per barrel, and it could have fallen to much lower levels had that situation continued simply because of the huge amount of idle production capacity in OPEC and the very low cost of producing its oil. For some time, the netback value pricing formula explained earlier became, in fact, the order of the market for the entire oil trade, so that most OPEC crude was moving through refineries rather than direct sales. This necessarily meant an increase in world demand for OPEC oil, which rose from 15.4 mb/d in 1985 to more than 18.3 mb/d in 1986. This average increase of 19 percent happened in spite of the fact that, from September to December 1986, OPEC had returned to its self-imposed production restraint.

It was not only the low oil price that led to such an increase in world demand for OPEC oil. The market-oriented pricing formula, notably the netback, also guaranteed a reasonable and permanent profit margin to refiners, who took OPEC oil not only to meet growing consumption and trade requirements but also to benefit from the situation by building up huge stocks of cheap oil.

Market Share Strategy Fosters Competition Within and Outside OPEC

The increase in OPEC production in 1986 was, however, unevenly distributed among its members, with large-reserve countries benefiting from the new situation much more than others. In one month (for example, August 1986, when the competitive situation peaked), Saudi Arabia's production reached 6.2 mb/d, or three times as much as the low levels reached in the summer of 1985, while the production of most other OPEC members did not rise except marginally and, in many cases, was similar to that of 1985, if not less. Table 1.3 compares individual member countries' production during the third quarter of 1986 (during which OPEC's total production peaked at about 19.4 mb/d) with the same period in 1985. Countries like Algeria, Ecuador, Gabon, Indonesia, Libya, and Qatar were unable to expand their production in such a way as to partially offset heavy losses in per barrel income, resulting from the prior decline, because of either limited production capacity, as in the case of Algeria, or marketing difficulties. In some other cases, like that of Iran, OPEC member countries could not even produce their quotas. The countries that benefited from the market share strategy, in addition to Saudi Arabia (a 132 percent increase in production), were Kuwait (up 69 percent), Iraq (up 50 percent), and, to a lesser extent, Nigeria and Venezuela.

Obviously, such uneven volume gains from the new pricing strategy had to create tensions within OPEC. The price crash of 1986 cost the organization about $55 billion in oil revenues; but the impact of that loss was as unevenly distributed among member countries as the gain in export volumes, as shown in Table 1.4. Given the production/marketing structure of Saudi Arabia's oil and its enormous idle pro-

Table 1.3. OPEC Crude Oil Production (in thousands of barrels per day)

	Third Quarter 1985	*Third Quarter 1986*	*% Change*
Algeria	676.2	680.4	0.6
Ecuador	28.5	217.3	−24.1
Gabon	166.3	162.6	−2.2
Indonesia	1,188.6	1,272.1	7.0
Iran	2,220.8	2,023.0	−8.9
Iraq	1,329.8	1,982.9	49.1
Kuwait	831.9	1,404.1	68.8
Libya	1,003.9	1,097.8	9.3
Nigeria	1,264.1	1,517.2	20.0
Qatar	305.2	359.7	17.8
Saudi Arabia	2,346.2	5,436.5	131.7
United Arab Emirates	1,041.1	1,435.5	37.9
Venezuela	1,563.3	1,875.1	19.9
Total OPEC	14,224.0	19,464.0	36.8

Source: Direct communications to the OPEC secretariat whenever available, otherwise estimates based on secondary sources.

Table 1.4. Distribution of the Loss of Revenue among OPEC Member Countries

	Dollar Income (billions)		%
	in 1985	*in 1986*	*Drop*
Algeria	9,170	3,760	59
Ecuador	1,927	983	49
Gabon	1,668	848	49
Indonesia	9,083	5,451	40
Iran	13,115	6,600	50
Iraq	11,380	6,980	39
Kuwait	9,729	6,200	36
Libya	10,520	4,700	55
Nigeria	12,338	6,300	49
Qatar	3,355	1,460	56
Saudi Arabia	25,936	21,190	18
United Arab Emirates	13,395	5,890	56
Venezuela	10,352	6,713	35
Total OPEC	131,967	77,073	42

Source: OPEC Annual Statistical Bulletin (1986).

duction capacity, the Saudis were able to move far larger quantities in the market than other producers. Hence they were the least adversely affected by such a strategy (18 percent reduction in revenue) compared with other countries (as high as 59 percent for Algeria, for example, and 42 percent average for OPEC as a whole). The limited production capacities of many countries did not allow them to compensate for the loss due to the price drop with a gain through volume. Obviously this situation led to the growing political tensions within OPEC. The low-reserve countries, which constituted the numerical majority of the organization's members, found themselves to be the real losers from the new strategy. The high-reserve countries, on the other hand, looked more to the future benefit of low price in terms of market share than the short-term losses in terms of revenues which were in their case less dramatic than in the first group.

Effect of Market Share Strategy on Non-OPEC Production

More importantly, the so-called market share or price war strategy played havoc with the oil industry outside OPEC, not only in such high-cost producing areas as the United States and the United Kingdom but also in Third World oil-producing countries. These countries found themselves, in one way or another, totally unable to face the new situation. Especially hard hit were those nations with high external indebtedness, such as Mexico and Egypt.

In the United States, production fell from 9.1 mb/d in January 1986 to 8.3 mb/d in September 1986; the 800,000 b/d drop came mainly from the so-called stripper wells, where the per barrel cost was too high to put this oil at a competitive advantage vis-à-vis the low-cost imported oil.[7] The price crash led to a recession in the country's oil-producing states: Oil investments were curbed drastically; the number of operating rigs was reduced to less than one-fifth of their peak of 1983; oil-related loans created a real threat to the banking system, etc. Although in other areas producers continued to sell at the low price because it was still sufficient to cover their running/operational costs, the investments needed to sustain high production levels from these areas in the medium and longer term were in real jeopardy, with most of the companies' investment programs being cut by more than half.

Had OPEC continued this strategy of gaining a higher market share, production from the United States and the North Sea would have dwindled dramatically in a matter of two years. Figures indicate clearly that the rise in OPEC production during the third quarter of 1986 (when OPEC output was more than 5 mb/d higher than a year earlier) meant that the organization in its entirety had been able to win the volume battle, especially if we consider the fact that this very substantial increase was effected in only two months of the quarter, since OPEC had to return to its production restraint on 1 September 1986.

The increase in OPEC's share of the world market, however, created another source of tension with the industrialized countries, especially the United States, which for some time was seriously considering the imposition of import tariffs to protect its high-cost oil. Moreover, the OPEC oil confrontation was not limited to the industrialized countries but also concerned other Third World oil-producing countries, notably Mexico, Egypt, Oman, and Malaysia. These countries, seeing a real threat to their development prospects in OPEC's new strategy, hastened to express their willingness to come and talk to the organization, in contrast with their earlier attitude when they had turned a deaf ear to the organization's repeated appeals for cooperation. With the new situation, these countries suddenly found themselves not as free to compete with OPEC as in the past, when it had been sufficient for them to undercut the organization's fixed price in order to move greater volumes. This is why in 1986, especially during the second and third quarters, we find production in most of these countries falling to levels lower than in 1985. Egypt's production, for example, was reduced by more than 15 percent during the summer of 1986, compared with the average for 1985; the same applied to Mexico, whose production in the summer of 1986 was 10 percent less than it had been in 1985.

The confrontational aspect of the so-called OPEC price war strategy went beyond oil-producing countries to affect consuming countries. After the initial jubilation at the price crash, they awoke to the fact that low oil prices could jeopardize their investment programs in energy diversification and could well lead to greater dependence on imported oil, an eventuality they had been fighting against for the previous ten years. The benefits of the lower cost of oil imports could well be offset by the accompanying slowdown, or even halt, in the drive toward more energy conservation and diversification for their national security of supply.

The three-dimensional pressure on OPEC (i.e., the mounting tensions within the organization itself, the growing fears in industrialized oil-producing countries about the adverse impact of the new OPEC strategy on their energy and oil situation and industry, and from the oil-producing developing countries) led to a sudden halt in OPEC's drive for market share. Politically, the organization was too weak to sustain a confrontation of that kind on such a wide range of fronts. The first reversal of the situation was OPEC's decision to return to production restraint. As of 1 September 1986, it adopted an interim arrangement whereby its ceiling was fixed once more at around 16.2 mb/d (excluding Iraq) while prices were left to be determined by the market. As a result, prices increased to a range of $13 to $15 per barrel during the period September to December 1986. Finally, the coup de grace to OPEC's market share strategy was finally made by the main potential beneficiary of that strategy, Saudi Arabia. Toward the end of 1986, in a political gesture of solidarity with OPEC, Saudi Arabia announced its new policy of reverting to the fixed price system instead of to the market-oriented one which had prevailed during that year. Consequently, OPEC concluded its December 1986 agreement, by which it readopted the fixed price system at a level of $18 per barrel for a basket of seven OPEC crudes as mentioned earlier. That major policy change in OPEC echoed the mounting pressures from within the organization and outside to restore price stability in the marketplace.

The 1987 Precarious Stability

In concluding the December 1986 OPEC agreement package on price and production for 1987, stability around $18 per barrel price was in fact restored, but only temporarily.[8] The re-established fixed price system, although politically motivated by Saudi Arabia and Iran, who co-sponsored the package, did not change the market fundamentals which caused the price collapse of 1986.

Although revitalized by OPEC lower prices, the demand for OPEC oil did not grow in 1987 to the extent of solving the OPEC member countries' quota system. In 1987, world oil demand grew by about 2 percent. However, this increase in world demand, and hence the increase in the call on OPEC oil, was partially offset by an addition to supplies outside OPEC, which in 1987 was 40 mb/d, the outcome of investments in non-OPEC oil made when prices were high. The increase in the call on OPEC oil was therefore too small to solve the basic problems of member countries, especially those of Iraq, which in 1987 had substantially increased its export outlets, reaching a total of around 2 mb/d that year. UAE, on the other hand, persisted in its continued breach of OPEC's agreement regarding its quota so that its production was exceeding its quota by over 50 percent. This is apart from the quota problems of small producers like Qatar and Ecuador. By the summer of 1987, the ghost of price collapse was still lingering on the OPEC scene. Furthermore, in reverting to a fixed price system, OPEC partially absolved the non-OPEC producers of their responsibility for co-shouldering the burden of price defense through cuts in their production. Non-OPEC production increased in 1987 over 1986, although not substantially (the increase was partially offset by the fall in U.S. production). Mexico kept its promise of reducing its total production by 7 percent below capacity, and Norway informed OPEC of its reduced production by 7 percent below its new planned capacity.[9]

More importantly, a radical change happened in the Saudi policy concerning the price. That country's policy makers have come to realize, perhaps too late, that in a depressed demand the price of oil and the export volumes cannot both be defended at the same time. The kingdom had to choose between defending prices at a fixed level by sacrificing volume or defending the country's share of the market by abandoning the fixed price system and adopting one formula of market-oriented price mechanism or another. Selling strictly at the official OPEC price would mean automatically a "swingdown" of Saudi Arabia's production, even if such a role was explicitly spelled out in the OPEC agreement or not. We have already explained how, in abiding by the fixed price system for it's oil, Saudi Arabia's marketing pattern would not provide it with such marketing flexibility to sell all its quota. In a state of oversupply the fixed pricing system would simply mean that the Saudi production (or for that matter that of any crude oil producer offering in the market its crude at no less than the official selling price) would have to be less than its quota. Buyers would lift first cheaper oil from other OPEC countries before they came to the Saudis to lift such volumes of oil at a fixed price as to meet their requirements. Few OPEC spokesmen were aware of the fact that the fixed price system reduced OPEC to a swing producer of world energy supplies in order to support the structure of the world oil price, and within OPEC, Saudi Arabia had to be the swing producer to support the OPEC price structure—a curious price structure which was doomed to collapse. For this reason, Saudi Arabia had to face soon the same problems that it faced in 1985 by agreeing to a reduction in its production, although this time to a much lesser extent. In fact, Saudi production in the first quarter of 1987 was more than 600,000 b/d short of the then quota for the country of 4.35 mb/d. During the second quarter, Saudi Arabia had again to swing down in order to save the price and, in general, market stability during the first half of that year. Without this swingdown, prices would have fallen again, but not of course to the low levels of 1986. Naturally such a situation could not continue as the Saudi government was not willing to lose any number of barrels of its quota. All those developments have provoked, as of the latter part of 1987 and all of 1988, a shift to a market-oriented pricing system by which all the member countries were able to market their entire quota. This shift in price system was made easier by the growing demand for OPEC oil, as we shall see later.

With the shift of Saudi pricing policies to the virtually same pricing system of 1986, the fixed price system died a natural death. Market realities were far stronger than the country's decision taken under political pressure from within and outside OPEC.[10] In fact, no OPEC country in the entire year of 1988 was selling crude oil at the official selling price. All member countries were adopting one competitive system of pricing or another that enabled them to market their entire quota, ranging from the netback value system to barter deals, processing agreements, counter-trade, or financial settlements between OPEC selling countries and buyers.

OPEC seems to have finally admitted the lack of realism that characterized its former oil pricing systems and that had cost the organization highly in terms of market share. Fixed prices cannot be sustained against changing market realities. The agreement of November 1988 that formally dropped the fixed price system could in a way be considered a turning point in the OPEC approach to the problem of prices. The new OPEC realism manifests itself in that agreement by which the de facto abandonment of a fixed price system was formalized. Prices are no longer fixed at any level and member countries of OPEC are no longer committed to any price. According to the agreement, the price of $18 per barrel is no more than a target price that could or could not be reached through production regulation. This meant that each OPEC country has the freedom to sell its quota the way it deems in its best interest. With this universally accepted, market-oriented pricing system, the only guarantee to price stability that OPEC could offer to the world was the self-imposed restraints on production, that is, the quota system. The more OPEC countries abide by their quotas, the higher market prices and the nearer OPEC comes to the organization's target price of $18 per barrel.

OPEC's commitment to price stability in oil only through the regulation of its total production so as to match demand (the call on OPEC oil) marks a spirit of realism that could have a far-reaching positive impact on the world energy and oil situation. By leaving the pricing of oil, including the definition of price differentials for OPEC's crudes (the values of each OPEC crude in relation to the others), to each member country to deal with in the light of market forces, OPEC has demolished a big psychological barrier that was separating it from market realities.

Besides market-oriented pricing mechanisms, the November 1988 agreement marks another step toward realism in dealing with oil affairs: the inclusion of Iraq in OPEC production agreements by accepting this country's demand for quota parity with Iran. As of January 1989 Iraq had to abide by a production level which corresponds to that of Iran. By this agreement, that country became accountable for its production after practically four years of noncommitment to any production restraint.[11]

The 1986 Price Collapse Reversed Market Trends

The new OPEC low price profile, triggered by the price collapse in 1986, gave rise to changes in the world energy balance that helped OPEC toward the greater capacity of price administration (Table 1.4). The correction of past mistakes in OPEC's price policies and structure toward more realism, mainly the change in pricing from a fixed price system to market-oriented prices, has strengthened OPEC and its capacity to administer prices in the world market. This change by itself has strengthened OPEC by removing the conflicting situations in marketing patterns of OPEC countries so that no single country or group of countries within OPEC would take the task of defending price alone. With the new price formula, there is a greater possibility of more collective price defense mechanisms through proportionate production regulation. More important, low price policies triggered a series of reversed structural changes in the oil industry during the last four years or so. Low prices have revitalized growth in world demand for oil, which has been positively responding to lower prices and to higher-level economic performance in industrialized and semi-industrialized countries.

Between 1985 and 1989, world consumption of oil has risen from about 46 mb/d to over 52 mb/d, or an increase of about 3 percent per annum. In Organization for Economic Cooperation and Development (OECD) countries, where conservation measures have been very efficiently implemented, consumption of oil rose from a low of 34 mb/d to over 37 mb/d during those four years. In developing countries also, especially those with high rates of industrialization such as Brazil, India, and South Korea, consumption has been growing at an even higher rate, exceeding, in certain cases, 5 percent per annum.

Lower oil prices helped the world economy to grow faster in the same way that higher oil prices in the 1970s and 1980s caused or at least contributed to the slowdown of the industrialized countries' economic performance. Naturally, higher levels of economic prosperity would lead to more consumption of energy. Lower prices could activate demand through what has turned out to be higher income elasticity of demand for oil, which has risen from a low level of .5 percent during the 1980s to reach almost the same level of the 1970s (i.e., unity), so that a marginally less than 1 percent increase in the GNP is coupled with a marginally less than 1 percent growth in oil consumption. It could be generally thought that the decoupling of energy consumption from economic growth which was achieved in the 1980s as a result of energy conservation is replaced by energy recoupling, so that higher economic growth would necessarily be accompanied by similarly high consumption rates. It is thought in this context that after peaking in the late 1970s and early 1980s, energy conservation has been relaxing.

On the supply side, oil prices have had the effect of slowing down the growth rates of supplies outside OPEC. Expensive U.S. production has not been able to compete with cheaper imported oil from the Middle East. In order to maintain production levels in the United States, huge capital investments are required. The result was that from a peak of 9 mb/d in 1984, U.S. production went down by about 1 mb/d in 1988 and the expected U.S. production in 1989 would still be reduced to about 7.5 mb/d. With U.K. production reaching a plateau at more than 2 mb/d and the availability of Norwegian oil, thought not to exceed 1.5 mb/d for many years, OECD production is failing to meet rising international demand. On the other hand, production from developing countries has been slowing down enormously. The oil supply phenomenon of important oil discoveries like those in Angola, Egypt, Oman, and others are not to be repeated. New additional capacities coming from the Yemen and Syria are rather small. Over the last few years, those additions are not reaching more than 200,000 b/d. With rising world demand for oil and stagnating if not decreasing supplies of oil outside OPEC, the call on the organization's oil has been growing very fast over the last few years. OPEC 1989 production is estimated now at over 22 mb/d compared with 15.5 mb/d in 1985, an increase of about 42 percent in four years.

Possible OPEC Future Price Scenarios

Those increases in OPEC production have enormously improved the organization's capability of administering price in the future. The higher the production, the easier it is to solve the quota problems. Future OPEC capability to administer prices will depend on how much the world call on its oil will increase, on the one hand, and how much OPEC countries themselves are ready to invest in production capacity in order to meet increasing demand for its oil. The fall in OPEC production for many years has had the effect of reducing the organization's production capacity from its peak of 34 mb/d in 1979 to about 27 mb/d today.[12] If the trends of world demand and non-OPEC supply of the past few years continue into the future, the call on OPEC may reach 26 mb/d, which is equal to the present capacity. If investments to maximize OPEC production capacity are not made by OPEC countries to increase production capacity significantly, price explosions cannot be avoided sometime by or after the mid-1990s. It is estimated that with necessary investments OPEC capacity by the year 2000 could reach as high as 33–34 mb/d.[13] Only such capacity could comfortably surpass world demand of that time. However, investments to increase OPEC production capacity would require enormous capital investments, which are not easily made by OPEC countries, most of which are heavily indebted. Small and moderate price increases are therefore required to provide funds for investment. Such price increases are also necessary to mildly discourage high growth rates of demand for OPEC oil in order to avoid a market disruption. If not controlled by OPEC, prices could fluctuate violently. Therefore, if the currently low OPEC price level is main-

price, although higher than in 1986, is still not high enough to motivate investments in the aging U.S. oil fields in the lower forty-eight states, and not only in the stripper wells.

8. The OPEC package for 1987 consisted of Iran's agreement, under OPEC pressure, to exempt Iraq from the quota agreement as the only way to solve the impasse, as Iraq did not recognize its previously allocated quota of 1.2 mb/d. Therefore, the new 14.8 mb/d ceiling comprised the national quotas of twelve member countries and not thirteen. The package also reiterated member countries' commitment to a fixed reference price on the basis of which other OPEC crudes were priced. Arabian Light (which in the past served as a marker crude for OPEC and which was included in the new pricing system of OPEC based on the basket of seven crudes) was priced at $17.52 per barrel.

9. The production cut by those two countries, especially Norway, was fictitious. Planned or even installed capacity is a theoretical concept that can or cannot be ascertained in reality except by producing at full capacity in a sustained manner throughout a certain period. It is more of a target capacity than an actual sustainable capacity that can or cannot be achieved in reality. Moreover, it is generally accepted that producing at full capacity does not necessarily mean 100 percent of capacity. Marginal idle capacity should always be kept to face operational and seasonal production variations.

10. Besides the mounting financial pressure inside Saudi Arabia, the Kingdom was under enormous political pressure from within OPEC to return to the fixed price system as the only means to stabilize the market prices at any level. Foremost among the OPEC members that exercised pressure were Algeria, Libya, Iran, and Venezuela. From outside OPEC there was pressure from producers like Egypt and Oman. Even producing countries from within the Organization for Economic Cooperation and Development (OECD) expressed dissatisfaction about the adverse impact of OPEC market share strategy on their production.

11. Prior to the Gulf War of September 1980, Iraq's increasing production reached a level of 3.8 mb/d. Most of that oil was exported via deep-water terminals in the Gulf, the other parts via the Turkish and Syrian pipelines. By the beginning of the war, all the Iraqi export installations in the South (the Gulf deep-water terminals) were destroyed by the Iranian artillery. Thereafter, the Syrian pipeline was closed by the Syrian government mainly for political reasons, as Syria was allied to Iran. Iraq production levels were constrained by the export outlets (only the Turkish line, with a limited capacity of 0.75 mb/d). Iran, on the other hand, which did not suffer from export constraints, was given in the 1983 agreement a quota twice as much as Iraq (2.4 mb/d). As Iraq started to produce more, it insisted that its quota should be equal to that of Iran. In the first instance the OPEC conference was not able to reach an agreement to satisfy Iraq, but this country was finally able to get quota parity in accordance with Iran in the November 1988 agreement.

12. From its peak of about 34 mb/d in 1979, OPEC production capacity had by the end of 1990 fallen to 26 to 27 mb/d, as follows: Saudi Arabia,

8 mb/d; Iraq, 3; Iran, 3; Kuwait, 2.5; UAE, 2.3; Qatar, 0.4 (total Gulf, 19); Venezuela, 2.5; Nigeria, 1.8; Libya, 1.4; Indonesia, 1.2; Algeria, 0.7; Ecuador/Gabon, 0.5. The reason for this decline is the persistent under-utilization of the capacity as a result of the falling demand for OPEC oil. Huge investments are required to restore the capacity back to and beyond what it was in 1979.

13. A total of 33 to 34 mb/d, or an additional 7 to 10 mb/d, would be coming mostly from Saudi Arabia, Iraq, Kuwait, and probably Abu Dhabi.

HOSSEIN ASKARI

2. Saudi Arabia's Oil Policy: Its Motivations and Impacts

DURING the mid to late 1970s, the increasing demand for OPEC oil resulted in OPEC's genuine miscalculation that higher prices were the answer to restoring demand-supply balance to the market. This policy in time, as noted by Chalabi in Chapter 1, resulted in a continuous decline in the demand for OPEC oil, the residual supplier. From 1981 to 1985, while OPEC was continuously losing market share, it did not significantly reduce prices. Saudi Arabia, by acting as a self-appointed swing producer, maintained prices by reducing its own output, which had substantially increased from historical levels due to the Iranian revolution and the Iran-Iraq war.[1] This was, however, becoming an increasingly untenable economic position for Saudi Arabia by mid-1985.

For a few days in August 1985, Saudi Arabia's oil production hit a low of 2 million barrels per day (mb/d), with no end in sight for the continuation of the decline. It reversed its policy of acting as a swing producer and began to sell oil on a netback basis. The result was a dramatic collapse of prices on the world market. As a result, oil prices fell from about $28 per barrel in December 1985 to a low of $6 during 1986. In December 1986 the Saudis agreed to terminate their policy of netback sales and to cooperate in re-establishing an OPEC quota. This policy reversal, coupled with OPEC's self-imposed production restraints of September 1986, resulted in roughly a doubling of prices and the restoration of market stability. The swings in oil prices over a period of roughly one year had been from $28 to $6 to $18.

With roughly 25 percent of the world's proven oil reserves, a maximum sustainable capacity of a little less than 10 mb/d (with some investment and a lead time of about six months), and the ability to increase maximum sustainable capacity, Saudi Arabia has significant market power in influencing oil prices. The impact of Saudi oil policy on prices is generally accepted, but the motivations that contribute to the making of oil policy are less clear. In this chapter, the motivations behind the important shifts in Saudi oil policy during 1985–86 are identified by examining the political and economic contexts in which they were made.

Motivations of Saudi Oil Policy

Oil policy in Saudi Arabia is determined at the highest level of government. The king, the crown prince (deputy prime minister), and the second deputy prime minister are the ultimate decision makers. Some input is solicited from the Higher Petroleum Council, composed of selected members of the cabinet. As a result, Saudi Arabia's oil policy has been historically motivated by broad political and economic factors. The major political considerations have been Saudi Arabia's role in the world, Arab solidarity, the Arab-Israeli conflict and regional politics, cohesiveness within OPEC and the Third World, and Western (especially U.S.) support. Economic factors include long-term diversification from oil and short-term revenue needs to meet both domestic and external financial needs and obligations. The most important political aspect of oil policy is to preserve Saudi Arabia's influence in regional and world affairs. This is most likely to be achieved if liquid fuels remain the critical element in energy supply and Saudi Arabia maintains its dominance in oil supplies. Price developments must, therefore, be consistent with these long-term objectives.

Arab solidarity and the Arab-Israeli conflict are closely related. In the aftermath of the October War and the helplessness of the Arab world, oil was seen as the only "weapon" available to demonstrate Arab anger at U.S. support of Israel. Saudi Arabia played a key leadership role in cutting back oil production by 5 percent a month. This oil as weapon concept, though less potent, is still alive today in the minds of many Arabs, and consequently Saudi Arabia is mindful of how its oil policy is perceived in the Arab world.

Although Saudi Arabia is stable and has a well-equipped military force, it still must take into account regional considerations, such as its two powerful neighbors, Iraq and Iran. Consequently, Saudi oil policy has incorporated the impact of this policy on these two combatants. Given that Iraq, and to a much greater degree Iran, are not as well endowed with oil on a per capita basis as is Saudi Arabia, a drop in the price of oil has a bigger effect on their economies.[2] And in the case of Iran, with its low production capacity and with war expenses which were paid in cash, it was imperative that each barrel of oil brought in as much revenue as possible.

Saudi Arabia's political influence in international affairs is derived from various sources. In part, it comes from its position as the protector of the two holiest cities in Islam, Mecca and Medina (the king is officially addressed as the "Custodian of the Two Holy Cities"). But Saudi Arabia's influence mainly flows from its economic/oil position. As the giant among oil producers, Saudi Arabia is important to the economic well-being of oil exporters and importers alike. Through its wealth, its ability to render assistance to various regional governments and causes (such as the Mujahedeen in Afghanistan) and to other Third World countries has given it significant influence. And in the late 1970s and early 1980s, the size of its financial reserves gave it an important voice in international financial markets. As a country that has such diverse interests, there are aspects of its policies that will be in conflict with one another. For example, to the extent that some oil exporters demand a higher price for oil and powerful oil importers desire a lower price, Saudi oil policies cannot please everyone. However, cohesiveness within OPEC and the interests of important Western countries, especially the United States, have been given priority.

Saudi oil policy is, of course, also motivated by important economic considerations. Saudi Arabia is truly an oil-based economy and, as such, its actual net national product (NNP) cannot correctly be compared to the NNP in industrial countries; that is, actual or conventionally measured, NNP in Saudi Arabia does not represent maximum, long-run, sustainable consumption, the theoretically correct basis for NNP. In a resource depletable economy, a larger portion of actual NNP must be saved and invested in order to generate alternative (sustainable) sources of income as oil is depleted.[3] In such a setting, Saudi Arabia's long-term interests are clear: to maximize discounted oil revenues (a technically difficult task but necessary to attempt anyway) and to save and invest a significant portion of oil revenues in profitable investments, foreign or domestic, to compensate for oil depletion.

These long-term economic interests at times conflict, however, with short-term needs. Given a high level of government expenditures, Saudi Arabia, in the short run, has high revenue needs, needs that must be met by oil exports since they have been unsuccessful in generating other sources of income. Although Saudi Arabia's net external assets have been considerable, exceeding $160 billion in 1981, they had fallen dramatically to about $85 billion by the end of 1985, and of this $85 billion, roughly $35 billion were in the form of loans to Third World countries, with Iraqi "loans" amounting to about $25 billion. Saudi Arabia's dependence on oil can be measured by various ratios. The ratio of oil exports to total exports has consistently exceeded 98 percent. Its dependence on oil revenues for government revenues has fluctuated (91.7 percent in 1980–81) and appears to have fallen (48.5 percent in 1985–86), but this is due to the fact that oil revenues have declined, making investment income, initially derived from oil, a more important factor. Although the share of oil-related gross domestic product (GDP) in total GDP may no longer appear as large (69.6 percent in 1980–81 to 34.5 percent in 1985–86), much of nonoil GDP is comprised of construction and sectors that are heavily subsidized, such as utilities and agriculture. Manufacturing contributes only 3.7 percent to GDP (1984–85)—excluding petroleum refining. Finally, as will be shown, short-run revenue needs are substantial, arising from two main expenditures: military and subsidies.

To the extent that Saudi Arabia's financial reserves are depleted or are expected to be depleted, Saudi oil policy is influenced by considerations that may maximize short-run revenues, as opposed to discounted long-run revenues. Politically, given the importance and size of military and subsidy expenditures, short-run needs become at times the overriding consideration.

In addition to domestic revenue requirements, Saudi Arabia has, or has had, substantial financial commitments to Arab neighbors and to other Islamic countries. Gulf Cooperation Council (GCC) member states (Saudi Arabia, United Arab Emirates [UAE], Bahrain, Kuwait, Qatar, and Oman), Iraq, Syria, Pakistan, the Mujahedeen in Afghanistan, and others receive substantial amounts of aid from the Kingdom. These financial commitments play an integral role in Saudi Arabia's political clout in its sphere of influence.

Finally, the availability of natural gas in Saudi Arabia is limited by oil production, since most of the gas in Saudi Arabia is in the form of associated gas. As such, the Kingdom needs some minimal amount of oil production to satisfy natural gas needs. To the extent that Saudi industry should become dependent on gas as a feedstock or as a fuel, satisfying this need may also become a factor in Saudi oil production policy. In the future, this problem may be somewhat ameliorated due to the discovery of unassociated gas in Saudi Arabia and the acquisition of gas from Qatar. During the last fifteen years, all these factors have come into play at one time or another, contributing in varying degrees to the formulation of Saudi oil policy.

Saudi Oil Policy: 1973 to the Present

1973–74 to 1978. In the aftermath of the 1973 October War, the Saudis were in the vanguard of the Arab oil embargo. At a meeting of the Organization of Arab Petroleum Exporting Countries (OAPEC) in Kuwait on 17 October 1973, it was decided to cut back on production by 5 percent a month until Israel evacuated all occupied territories, and at another meeting on 4 November these cutbacks were enlarged to 25 percent. In addition to the general cutback, there was a selective embargo against Western supporters of Israel and a concerted price rise of 70 percent instituted by all Arab countries.

The market at this time was perfect for the success of such actions: There had been a rapid expansion of demand for oil while at the same time there had been no comparable expansion of exploration for non-OPEC oil. OPEC was at this time producing each marginal barrel of oil. It was, however, Saudi Arabia's adherence to the agreed-upon cutbacks and price rises, coupled with panic buying of oil in importing countries, that was the catalyst for success of the embargo. At this time, the Saudis were in no great need for higher revenues. Indeed, their revenues at this time exceeded their financial commitments and requirements.

After the oil embargo and up to 1978, OPEC became a forum for the Arab world, and as a major player within OPEC, Saudi Arabia had become a regional and a limited world economic power. To support this position of newfound clout, Saudi Arabia wanted to maintain nominal prices in order to support OPEC. At the same time, the Saudis did not want the price of oil to rise too high for fear of jolting the world economy further and causing a backlash against OPEC, which would in turn damage the oil market, their source of income. During this period, internal Saudi economic considerations were not dominant, since revenues still outpaced expenditures.

1978 to 1981. During the period 1978 to 1981, first the Iranian revolution and then the Iran-Iraq war caused cutbacks in production from the two countries, and as the market anticipated the worst, there were surges in panic buying, which forced up the price of oil. In response, Saudi Arabia increased its output to the maximum sustainable capacity of nearly 10 mb/d from 1979 to 1981, thereby preventing even further price rises.[4] The Saudi rationale for putting the brakes on price rises was based on a combination of political and economic factors. It wanted to avoid a further shock to the world economy. Economically, a low price was seen to be in the best long-run interests of the Saudi economy by keeping the oil market healthy and intact.

1982 to 1985. From 1982 until late 1985, Saudi Arabia continued to act as a swing producer to maintain OPEC price levels. The high prices that had helped cause a worldwide slump in the demand for oil were effectively even higher when one takes into account the exchange rate of the dollar during these years. Oil prices are quoted in dollars by OPEC, so if the dollar rises in value relative to other currencies, then the effective price of oil rises for the world as a whole. Consequently, even after the dollar price of oil reached its peak (as it did in 1982), the price in other currencies continued to rise. In this case, the dollar gained strength on international markets until 1985, and with it went the price of oil, as measured by a composite of other currencies (see Table 2.1). This further reduced the demand for oil, and Saudi Arabia had to cut back production even further.

This policy was a clear mistake for Saudi Arabia. From 1974 through 1978, Saudi Arabia's oil production as a percent of total OPEC production averaged 27.24 percent (26 percent in 1974, 25.3 percent in 1975, 27.4 percent in 1976, 29.1 percent in 1977, and 27.5 percent in 1988). This was a period with no quotas, and each member of OPEC produced at its own desired level. Saudi Arabia had slightly increased its output relative to other OPEC countries during this period in order to maintain lower oil prices, which it perceived as being in its own national interest. From 1979 through 1981, Saudi Arabia's share increased dramatically (30.4 percent in 1979, 36.4 percent in 1980, and 42.7 percent in 1981) because of disruption in Iran and the war. Clearly Saudi Arabia had again acted in its own self-interest. In 1982, Saudi Arabia was unwilling to accept a formal quota and opted to act as a swing producer. This resulted in an even more dramatic decline in its share to 33.6 percent in 1982, 28.3 percent in 1983, 25.7 percent in 1984, and 20.9 percent in 1985—a share substantially below that prevailing during 1974 to 1978 and more importantly with no reversal in sight. Indeed, within the OPEC countries Saudi Arabia and other GCC members of OPEC were the only countries in the aggregate where excess production capacity exceeded oil production between 1982 and 1985. Because of Saudi Arabia's high reliance on oil exports, it experienced proportionally lower export and government revenues. Beginning in 1982, Saudi Arabia started to finance current account and budget deficits from earlier savings by drawing on external assets.

It is difficult to justify, with rational economic reasons, the Saudi policy of helping bolster a high price. To be fair, like many other analysts of the energy market, the Saudis did not foresee the severity of the worldwide drop in the demand for OPEC oil and felt that an upturn would soon occur. Thus they had felt that their revenue shortfalls would be short-lived. Saudi Arabia still had high reserves, although they were being depleted. In 1981, its estimated net assets were over $160 billion. Consequently, the Saudis' financial needs were not yet critical. There were also political motives of various sorts. The Saudis hoped that if they led the way, other oil-producing coun-

Table 2.1. Oil Price Indexes for Selected Oil-Importing Countries (price index per barrel in local currency)

Year		*Oil Price*											
Year		*($/Barrel)*	*Index*	*Belgium*	*Canada*	*France*	*Germany*	*Italy*	*Japan*	*S. Korea*	*The Netherlands*	*Spain*	*Composite Index**
1979		17.26	100	100	100	100	100	100	100	100	100	100	100
1980		28.67	166	166	166	165	165	171	172	207	165	177	169
1981		32.50	188	238	193	241	232	258	189	265	234	259	214
1982		33.47	194	302	204	300	257	316	220	293	258	317	248
1983		29.31	170	296	179	304	237	310	184	272	242	363	232
1984		28.47	165	325	182	339	256	349	179	275	264	395	242
1985		27.95	162	328	189	342	260	372	176	291	268	410	246
	Jan.	27.95	162	350	183	369	280	380	188	278	289	385	260
	Feb.	27.72	161	362	185	380	288	392	191	278	299	369	263
	Mar.	27.69	160	364	189	381	290	402	189	281	299	357	266
	Apr.	27.62	160	339	186	354	269	380	184	284	278	358	245
	May	26.93	156	333	183	347	265	373	181	280	274	358	243
	June	26.81	155	327	181	341	260	365	176	280	267	358	237
	July	26.97	156	313	180	326	249	360	172	282	256	375	232
	Aug.	27.38	159	306	184	318	241	357	172	289	248	389	238
	Sept.	27.61	160	313	187	326	248	366	173	295	255	405	237
	Oct.	27.79	161	294	188	305	232	346	158	297	239	413	231
	Nov.	27.98	162	290	190	301	229	342	151	298	236	405	227
	Dec.	27.90	162	283	192	292	222	334	150	298	228	413	219
1986	Jan.	27.53	160	272	192	281	213	320	154	293	219	416	213
	Feb.	16.50	96	156	115	161	122	183	81	175	126	259	121
	Mar.	15.19	88	139	105	144	109	163	72	161	112	240	110
	Apr.	11.03	64	101	76	108	79	120	51	117	82	164	82
	May	12.17	71	110	83	118	86	130	54	129	88	184	72
	June	11.40	66	103	78	111	81	120	51	121	83	172	84
	July	9.36	54	82	64	88	64	96	39	99	66	136	66
	Aug.	11.54	67	97	79	106	75	114	47	122	78	163	80
	Sept.	11.95	69	100	82	109	77	117	49	126	79	174	83
	Oct.	12.05	70	99	83	107	76	117	50	126	79	168	82
	Nov.	12.65	73	105	87	114	81	124	54	132	84	174	87
	Dec.	12.68	73	104	87	113	80	122	54	131	82	171	87

tries, both in and outside of OPEC, might follow the lead and also reduce output. This too was an unfounded assumption. The Saudis were concerned with OPEC cohesion: They feared that if they had increased or even maintained output, OPEC might have crumbled. Similarly, since they use oil policy to maintain support among countries in their region of influence, the concessions they made during this period of time served the needs of other OPEC members. They were also trying to placate the Iranians, who had sent high-level delegations to Saudi Arabia to plead for a higher price. Iran, with a greatly reduced production capacity, desperately needed the higher prices and the consequent higher short-term revenues to finance its war effort. This ambivalence toward the Iranians is characteristic of a Saudi policy that has always tried to get along with both Iraq and Iran. Even in spite of its massive financial support of Iraq's war effort, Saudi Arabia still tries to keep a dialogue open with the Iranians.

Although the role of swing producer was not in their long-term interests, the Saudis continued producing substantially below their capacity for four long years, in part because they are patient and attempted to persuade other OPEC members to cut back on their own oil production. In Saudi Arabia, decisions are reversed only slowly. The Saudis, it has been said, only make tough decisions when there is no other choice. That point came in 1985.

1985–86. By 1985, it was more than clear that the Saudi goal of persuading other oil producers to cut back on production was not working: Several OPEC members were producing in excess of their quota allocations, and non-OPEC exporters were paying little attention to Saudi Arabia's sacrifice. Saudi pleas within OPEC for production cutbacks went unheeded. This was an intolerable situation for the Saudis: Had they continued producing at this rate, the Kingdom would have been importing oil within two years! So they instituted the policy of net-back sales of their oil, which began a price war and led to the dramatic fall in oil prices of 1985–86. Although Saudi Arabia had increased output during the 1970s and especially in the aftermath of the Iranian revolution and the Iran-Iraq war to slow down price increases, this was the first time Saudi Arabia had ever used the oil weapon to reduce prices.[5]

The impetuses for this dramatic shift were again several. After such a long and costly period of production cutbacks, the Saudis found themselves faced with short-term revenue needs. The severity of the need for short-term revenues was and continues to be caused largely by subsidy, military, and foreign assistance expenditures. As can be seen from Tables 2.2 to 2.5 and Figure 2.1, subsidies for everything

Table 2.2. Total Government Subsidies (in millions of dollars)

	1975 (1395)	*1976 (1396)*	*1977 (1397)*	*1978 (1398)*	*1979 (1399)*	*1980 (1400)*	*1981 (1401)*	*1982 (1402)*	*1983 (1403)*	*1984 (1404)*	*Total*
Agriculture	64	71	94	147	221	427	716	1,207	1,466	1,947	6,361
Industry*	35	52	77	103	182	289	381	531	471	361	2,482
Water	166	136	140	160	195	447	1,808	2,416	2,545	2,405	10,418
Electricity	88	138	245	496	934	1,420	1,882	2,603	2,391	2,161	12,358
Housing*	66	243	325	468	900	1,481	1,959	2,725	2,525	2,041	12,733
Public Investment Fund*	72	140	214	310	517	896	1,334	2,002	1,792	1,353	8,631
Food†	411	170	200	241	431	902	1,464	1,303	458	727	6,307
Saudia†	7	32	48	47	70	95	253	118	0	0	670
Fuel†	191	313	421	662	1,420	3,771	5,615	6,784	7,669	6,642	33,488
Total	1,100	1,295	1,762	2,635	4,871	9,730	15,412	19,689	19,317	17,638	93,448

* Only capital subsidy. † Only operating subsidy.

Table 2.3. Total Government Subsidies as a Percentage of Oil Revenues

	1975 (1395)	*1976 (1396)*	*1977 (1397)*	*1978 (1398)*	*1979 (1399)*	*1980 (1400)*	*1981 (1401)*	*1982 (1402)*	*1983 (1403)*	*1984 (1404)*
Agriculture	0.2	0.2	0.3	0.4	0.5	0.5	0.7	1.7	4.0	6.3
Industry*	0.1	0.2	0.3	0.4	0.7	1.1	1.4	2.0	1.8	1.4
Water	0.6	0.5	0.5	0.6	0.7	1.6	6.8	9.2	9.8	9.5
Electricity	0.3	0.5	0.9	1.8	3.5	5.2	7.1	9.9	9.3	8.6
Housing*	0.3	0.9	1.3	1.7	3.4	5.5	7.4	10.4	9.8	8.1
Public Investment Fund*	0.3	0.5	0.8	1.1	1.9	3.3	5.0	7.6	6.9	5.4
Food†	1.6	0.7	0.8	0.9	1.6	3.3	5.5	5.0	1.8	2.9
Saudia†	0	0.1	0.2	0.2	0.3	0.4	1.0	0.4	0	0
Fuel†	0.7	1.2	1.6	2.4	5.3	13.9	21.2	25.8	29.7	26.3
Total	4.3	5.0	6.7	9.6	17.8	34.8	56.3	72.0	73.0	68.4

Figure 2.1. Saudi Arabian government subsidies (in billions of U.S. dollars).

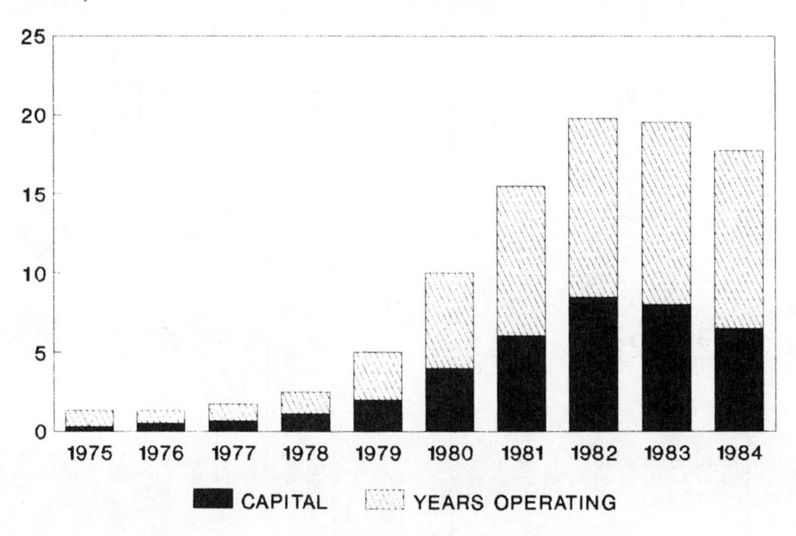

from water and electricity to food and housing have been large in size and have in the past taken up as much as 73 percent of oil revenues, accounting for nearly as much as 40 percent of GDP, and have been as high as $1,700 per capita on an annual basis. Military expenditures as a percentage of oil revenues have increased dramatically since the outbreak of the Iran-Iraq war (see Table 2.6) and have absorbed as much as 46 percent of total revenues. Over the past six years, Saudi Arabia has disbursed over $22 billion in official development assistance (or roughly 4 percent of GDP per year) primarily to Syria, Jordan, Bahrain, and Yemen.[6] Unfortunately, Saudi Arabia's subsidy policies had not resulted in significant nonoil export revenues, largely due to the concentration of subsidies toward social, as opposed to productive, goals (see Table 2.7) and the inefficient implementation of subsidies. Simultaneously, the government did not initiate an income tax policy to diversify and increase its own revenue base. As a result, the private sector continued to rely on government assistance, while the government needed oil revenues to sustain economic activity and its various programs.

When external reserves began to decline more rapidly than envisaged under the budget, a general sense of panic emerged. Usable reserves (net of previously mentioned loans to Third World countries and loans to Iraq, which mushroomed and are unlikely to be repaid) declined from a high of $160 billion in 1981 to our estimated figure

Table 2.4. Total Government Subsidies as a Percentage of Government Expenditures

	1975 (1395)	*1976 (1396)*	*1977 (1397)*	*1978 (1398)*	*1979 (1399)*	*1980 (1400)*	*1981 (1401)*	*1982 (1402)*	*1983 (1403)*	*1984 (1404)*
Agriculture	0.3	0.2	0.2	0.3	0.4	0.6	0.9	1.7	2.3	3.2
Industry*	0.2	0.2	0.3	0.4	0.8	1.2	1.6	2.2	2.0	1.6
Water	0.7	0.6	0.6	0.6	0.8	1.8	7.5	10.1	10.9	10.5
Electricity	0.4	0.6	1.0	2.0	3.8	5.8	7.9	10.9	10.2	9.4
Housing*	0.3	1.0	1.4	1.9	3.7	6.0	8.2	11.4	10.8	8.9
Public Investment Fund*	0.3	0.6	0.9	1.3	2.1	3.6	5.6	8.4	7.7	5.9
Food†	1.8	0.7	0.9	1.0	1.8	3.7	6.1	5.5	2.0	3.2
Saudia†	0	0.1	0.2	0.2	0.3	0.4	1.1	0.5	0	0
Fuel†	0.8	1.3	1.8	2.7	5.8	15.3	23.4	28.5	32.8	29.0
Total	4.7	5.5	7.4	10.4	19.5	38.4	62.2	79.3	78.6	71.8

* Only capital subsidy. † Only operating subsidy.

Table 2.5. Total Government Subsidies Per Capita (in dollars)

	1975 (1395)	*1976 (1396)*	*1977 (1397)*	*1978 (1398)*	*1979 (1399)*	*1980 (1400)*	*1981 (1401)*	*1982 (1402)*	*1983 (1403)*	*1984 (1404)*
Agriculture	8	9	11	16	24	44	70	113	132	170
Industry*	5	6	9	12	19	29	37	50	42	32
Water	22	17	16	18	21	46	177	227	230	210
Electricity	12	17	29	56	100	145	184	244	216	189
Housing*	9	30	38	52	96	151	191	256	228	178
Public Investment Fund*	9	17	25	35	55	91	130	188	162	118
Food†	54	21	24	27	46	92	143	122	41	64
Saudia†	1	4	6	5	8	10	25	11	0	0
Fuel†	25	39	50	74	152	385	549	636	692	580
Total	136	152	197	279	496	949	1,437	1,734	1,610	1,370

Table 2.6. Saudi Defense and Security Expenditure in Relation to Government Budget (in millions of Saudi Arabian riyals)

	1980	*1981*	*1982*	*1983*	*1984*	*1985*	*1986*
Defense and Security as percentage of							
Total expenditures*	23.1	23.0	27.2	27.7	31.4	33.3	35.7
Oil revenues†	17.1	19.9	35.8	44.0	65.4	94.7	NA
Total revenues	15.7	17.8	27.0	30.9	39.6	46.0	NA

Source: Ministry of Finance and National Economy.
* Budget figures.
† Excludes transfers from the petroleum sector.

Table 2.7. Classification of Subsidies by Overall Objective as Percentage of Total Subsidies for 1975–1984

Classification Subsidies	*Largely Social Objectives with Little Impact on Development of Competitive Industries*	*Mixture of Objectives*	*Largely Productive Objectives*
Electricity	7.4		3.2
Water	12.2		
Fuels		29.6	
Agriculture	5.5		
Food subsidies	6.6		
Saudi (operating)	0.7		
REDF	18.6		
PIF			12.5
SIDF			3.7
Total	51.0	29.6	19.4

Abbreviations: REDF, Real Estate Development Fund; PIF, Personal Investment Fund; SIDF, Saudi Industrial Development Fund.

of about $50 billion at the end of 1985. The Saudis knew that a plunge in the price of oil would not cost them significantly in short-term revenues. Whatever was lost in price would be made up in increased sales due to their capacity to produce more. An increase in production would also provide Saudi Arabia with some of its lost market share. This was not true for most other OPEC countries, since their production capacity was restricted. Hence the punitive aspect of the policy becomes apparent (for the impact of this policy on various OPEC members, see Chapter 1).

The Saudis felt that in the medium run, OPEC countries would see the loss generated by an oil price war and would reduce output. They also felt that non-OPEC countries such as Great Britain, Norway, and Mexico would cooperate by cutting their production. In

1986, OPEC and non-OPEC countries alike did recognize the implications for their own economies of an all-out price war. Realizing that they had more to lose than Saudi Arabia, some of these countries voiced their willingness to make some concessions. Within OPEC, for example, Nigeria agreed to refrain from cheating on its production quotas. Outside of OPEC, Egypt, Norway, and Mexico cut back on their production, although marginally.

In August 1986, after the price of oil had dipped below $8 per barrel, OPEC reached what might be considered a temporary agreement which set an aggregate quota of roughly 16.8 mb/d. This agreement was made possible by the actions of other OPEC members, especially Iran's suggestion that Iraq be excluded from any output limitations. As a result of the agreement, prices began to climb. This fragile agreement was extended at the October meeting and confirmed by Saudi Arabia in December of the same year. But an important question remained: While the agreement may have been helpful in the short run, was it in the long-run interests of Saudi Arabia?

In the short run, it was not a costly move for Saudi Arabia, since there was a small revenue decline. After August, by forcing other OPEC members to curtail cheating, the higher level of output coupled with a higher price (than in early 1986) increased Saudi oil revenues. It was in their medium- to long-term interests as well. If one examines two extreme scenarios, that of increasing output in pace with demand, or (again in the face of increasing demand) allowing prices to rise by keeping production levels constant, it becomes apparent that in the medium and long term, it was in Saudi Arabia's interest to increase production in pace with increases in demand.[7] This was true simply because Saudi Arabia had the lion's share of excess capacity, and the revenues generated by added production would be greater than those generated by smaller sales at higher prices. Such a scenario would avoid inducing exploration and conservation, again against long-range Saudi interests. Granted, such scenarios are extreme. But they clearly indicate the direction of Saudi economic interests.

Summary and a View to the Future

It should be clear that oil policies followed in the past by Saudi Arabia have not always been in accord with its own long-term economic interests. Through self-imposed production constraints, the Saudis after 1981 helped maintain a level of production that was too low, resulting in an inappropriately high price, a price that was especially high for many nondollar currencies as the dollar started a period of long and sharp appreciation. This high price level induced conservation, the development of alternative energy sources, and increased exploration and production among non-OPEC countries.

Forgoing long-run market share by following a production policy that sustained high prices, Saudi Arabia was forced to continually reduce production in order to maintain prices. The abandonment of the swing producer strategy in 1985 appears to have been to the benefit of Saudi Arabia and the GCC countries, both economically and politically. And if anything, the Saudis could produce even more in order to regain a larger share of the market, also resulting in higher medium- and long-run revenues.

Importantly, the Saudis have learned a hard lesson that market share and oil revenues are more in their interests than higher prices and the placation of other OPEC countries. Their high domestic expenditures (social, military, and foreign assistance) will force them to consider short- and medium-term revenue requirements, and since keeping production levels higher satisfies these requirements as well as long-range Saudi interests, they will probably follow this path. There really are no other significant revenue sources: Petrochemicals, for example, an industry in which the Saudis have invested heavily, contribute about 5 percent to total revenues.

In December 1987, the Saudi government took measures to increase revenues, to borrow, and to reduce expenditures. The revenue-increasing measures included new and higher tariffs, a variety of new taxes, and new user charges for some services.[8] The government authorized domestic borrowing (through zero coupon bonds) of up to SR30 billion ($8 billion) and a few subsidies were reduced. All these revenues are, however, not enough.

The initiatives to increase revenues and to reduce expenditures are not significant. More drastic measures, such as a national income tax and major reductions in the all-important subsidies such as electricity, will be required to address the medium-term revenue requirements of the government. But Saudi Arabia is unlikely to resort to such options. Instead, it is likely to borrow from the private sector and on the international financial markets and await the expected improvement in the oil market in the mid-1990s.

If oil revenues are counted on, as opposed to more domestic expenditure cuts, then the direction of oil policy is clear. The state of the oil market, with its current excess capacity of roughly 15 mb/d, and future worldwide supply and development forecasts also signal the benefit for Saudi Arabia of expanding production. And if Saudi Arabia does act to increase production, then this policy will make a rapid rise in oil prices unlikely.

With Saudi Arabia's revenue requirements, the inability to raise nonoil revenues significantly and to cut expenditures more drastically, and its reduced level of financial reserves, financial and economic factors will continue to dominate Saudi decision making in the foreseeable future. And if economic benefits are the major factor, then Saudi oil policy will continue to move away from price rises in favor of stable prices through steady production increases.

Notes

The author is very grateful for the contributions of Charles H. Wilbanks and Babak Dastmaltschi.

1. See Chapter 1 for an excellent discussion of the problems, tensions, and distribution of sacrifice within OPEC.

2. Iraq has the second largest proven oil reserves in OPEC. Some oil company executives have said in private that Iraq may ultimately have even larger reserves than Saudi Arabia. But on a per capita basis, Saudi Arabia is ahead of Iraq, although it ranks behind Abu Dhabi, Kuwait, and Qatar.

3. A detailed mathematical discussion of this point is contained in Hossein Askari (with Babak Dastmaltschi), *Saudi Arabia: Oil and the Search for Development* (JAI Press, 1990), with a mathematical appendix on this point by Martin Weitzman.

4. British Petroleum Company, *Statistical Review of World Energy* (June 1986).

5. Saudi Arabia's inability to arrest the reduction in its oil exports was further hampered by marketing difficulties and inability, and/or unwillingness, to discount prices (see Chapter 1 for a detailed discussion of these points).

6. OECD Press Release, "Financial Resources for Developing Countries: 1986 and Recent Trends," 19 June 1987, p. 5. For the sake of comparison, U.S. ODA contributions equaled only 0.2 percent of GNP in 1986.

7. This has been clearly pointed out in Abdullah El-Kuwaiz, "OPEC and the International Oil Market: Age of Realism," *OPEC Review* (Winter 1986). Chalabi in Chapter 1 makes a similar argument by showing that Saudi production increased dramatically relative to the OPEC average.

8. An income tax for expatriate workers was announced but almost immediately withdrawn. Higher user fees and taxes for a variety of services were also announced, but again some of these (roughly 35 percent) were almost immediately rescinded. It is estimated that the new revenue-increasing measures may raise between SR6 to 7 billion, or about 5 percent of government revenues.

EDWARD N. KRAPELS

3. The Fundamentals of the World Oil Market of the 1980s

THE fundamentals of world oil include not only the supply and demand for its products but also the political relations between the governments with large interests in oil: OPEC and the importing nations. Moreover, since the establishment of petroleum futures markets, perceptions about oil prices and supplies can also be said to be part of the fundamentals, especially since the futures markets give anyone with money and an interest in oil a vehicle for gambling on those perceptions and thereby for influencing the price of oil.

Often, analysis of oil affairs focuses only on the supply/demand fundamentals, or the politics between OPEC nations, or the actions of oil companies. Indeed, the analysis of international oil affairs has for decades been dominated by several images. The first is of the oil industry as a conspiracy. John Blair's *The Control of Oil* is the leading example of an analysis based on that image. The image is grounded in a view of the industry, most often associated with Paul Frankel's *Essentials of Petroleum* published in 1946, that sees the industry as inherently noncompetitive due to its remorseless need for high up-front capital investment.[1]

Blair took this as a starting point. His 1976 *The Control of Oil* revealed in striking detail the mechanisms—many of them collusive under American antitrust law—whereby the major oil companies had sought to influence oil affairs. The strength of Blair's book was its detailed treatment of these mechanisms; the shortcoming was its neglect of how the steadily advancing demand and the steadily declining share of supply of non-OPEC producers in the 1960s and 1970s were bringing the era of the majors to an end regardless of the sophistication of their mechanisms. Equally serious was Blair's neglect of government interests and actions: Companies did what they did not only because governments let them; they positively encouraged and sometimes forced companies to behave collusively. The establishment of the Iranian Consortium is the clearest example.[2]

The mirror image of the oil industry as cartel is the oil industry as inherently competitive, an image portrayed in Morris Adelman's landmark *The World Oil Market*, published in 1971. Adelman's book remains a tour de force of economic analysis and description, but like Blair's book its terms of reference are pre-OPEC. In direct contrast to Frankel, who had argued that the oil industry was inherently oligopolistic, Adelman argued that the industry was inherently competitive and that in the absence of government interference (that of the U.S. government in particular) the oil industry would not be able to organize an effective cartel. The experience of the 1980s might lend support to his argument, but it is not at all clear that the absence of government interference was the sole or even the principal condition leading to the emergence of the more competitive oil market of the 1980s. The deregulation of oil went further in the United States than almost anywhere else: In the majority of countries of the world, regulation of various degrees of stringency and effectiveness is still very much the rule.

The second set of images that has dominated thinking about oil is of OPEC as a cartel and its mirror image, OPEC as a failed cartel. There can be no question, or course, that the OPEC group has been important in international oil affairs since 1973. Much of the best work on OPEC, Griffin and Teece's *OPEC Behavior and World Oil Prices* in the realm of economics and Skeet's *OPEC: Twenty Five Years of Prices and Politics* in the realm of "contemporary histories," makes valuable contributions to the study of change in oil affairs.[3] These OPEC-centered analyses remedy the deficiency of Blair's book—they describe government interests and strategies—but they do not deal with the business of oil as thoroughly as Blair, Frankel, and Adelman did.

OPEC-centered analyses focus on the political process that yields, or fails to yield, decisions on the control over the marginal supply of oil. A more general analysis of control over oil must also examine the underlying supply-demand imbalance that OPEC has to redress. Why do such imbalances exist? The answers are in the demand side of the business, on how the structure of the business has led to the development of surplus production capacity, and on how effective are the industry's mechanisms (including OPEC meetings) at containing that surplus or preventing the holders of it from flooding the chronically glutted market with oil.

In spite of the restriction of their focus to the marginal supply piece of the overall control-over-oil equation, the OPEC-centered analyses have provided useful ways of examining how the member countries have governed (or failed to govern) the marginal supply of oil in pursuit of certain price or revenue targets. The central question about OPEC has been whether the forces that tend to drive the members apart (from simple greed to complex geopolitical differences) would be brought under sufficient control to allow the members to produce a workable arrangement on the volume and price of their oil sales.

Analysts have employed variations of cartel models to study this question. The swing producer model has been most commonly employed. In this way of looking at OPEC, the members are typically divided into two camps. In one camp are the countries that cannot or will not adjust production, in the other the countries that can and will adjust production. Much of the OPEC analysis has focused on who would be in what camp. Once that was analytically determined, the consequences of various groupings could be considered: The more countries in the swing producer group, the greater OPEC's influence over oil affairs.

This kind of analysis was most common in the late 1970s and the early 1980s, when many OPEC members appeared to have the luxury of choice between participating in or abstaining from the swing producer role. During the years that many OPEC members were earning more than they could spend (roughly, 1974 to 1983), speculation about the consequences of their being "spenders" or "savers," whether they had high versus low "discount rates," provided useful insights into their behavior. If OPEC's wealthy countries turned from savers into spenders, if they chose to govern their affairs as if they had high discount rates (i.e., preferred earning money now to earning it later), then the governance of the marginal supply of oil might suddenly change. What is difficult, of course, is understanding why countries change from one preference to another. As Adelman has noted, all economists can do is analyze the economic consequences of the preferences. Who really knows how sovereign states will finally behave?

Charles Doran's 1977 analysis of OPEC as a cartel provides the framework for one type of explanation of OPEC behavior.[4] Given that during the 1979–80 crisis the distribution of market shares and the level of oil prices were determined, Saudi Arabia was the only country with the revenue and production basis to be the price leader and cartel enforcer. Its leaders must have believed at the beginning of the 1981 to 1985 period that the $32 price was the optimal cartel price and the optimal price for Saudi Arabia. That is, at $32 per barrel, neither OPEC nor Saudi Arabia would lose enough market share to cause them to want to change prices and sales volumes. In effect, this belief was rooted in the assumption that global demand for oil was extremely inelastic to price. As long as that assumption held true, as Doran argued, cartel cohesion would remain high and OPEC would *administer* prices at $32 per barrel, rather than let the market determine prices, and that very act of administration was conducive to OPEC's unity and cohesion.

The reality of a high or even normal elasticity of demand (and of non-OPEC energy supply) to the very high prices set by OPEC in 1979–80, however, raised the cost of cartel leadership enormously.

Doran, Adelman, Griffin, and other analysts of OPEC behavior argued in the late 1970s and early 1980s that OPEC would have price-management problems if demand for its oil declined sharply. Adelman, for example, warned of the difficulties OPEC would have if it moved from operating like a "rational monopolist" model, with Saudi Arabia acting like the rational monopolist, to the "all the nations get together" model, where they would all have to take action to maintain a given price level. In fact, ever since Saudi Arabia has abandoned the dominant producer role, OPEC has had great difficulty carrying out its functions as a price-determining cartel.

These kinds of critical analyses of OPEC's strengths and weaknesses are an important component of the analysis of the 1980s, but they are only a component of what must be a broader survey of variables that determine the control over oil. OPEC behavior, it is argued here, is in the final analysis only the study of one of the *mechanisms* to bring international oil supply into line with demand. Such mechanisms are only one of four sets of variables determining the degree to which control can be exerted over international oil affairs.

The broader analysis of change in oil affairs must include changes in the fundamentals: the structure of the business, the financial strength of contending parties, and the mechanisms whereby the industry strives to create a situation where it does not have to rely on price to balance supply and demand. During the 1970s, their collective force produced a startling series of changes whose net effect was an enormous increase in petroleum prices. In the 1980s, the fundamentals worked the other way: the politics, economics, and business structure and practices of oil combined to force the price back down, in real terms, to a level not far removed from what it was in 1973.

The Increase in Surplus Oil Production Capacity

Figure 3.1 provides a view of the level of OPEC oil production as contrasted with its estimated production capacity from 1979 to 1988.[5] As indicated in Figure 3.1, OPEC's oil production capacity was stretched to the limit in 1979 and 1980, but beginning in 1981 the demand for OPEC oil declined until by 1985 output was less than half of capacity. With the sharp drop in prices in 1986, output began to increase again, but as of 1988 it is still far below OPEC's 1979 production levels.

The amount of surplus oil production capacity hanging over the world market is, of course, dependent on the level of global demand and the capacity of global supply. In the 1980s, there was finally a reaction on the demand side to the sharp oil price increases of the

ure 3.1. OPEC production capacity (□) and output (+). *Abbreviation:* mb/d, millions barrels per day.

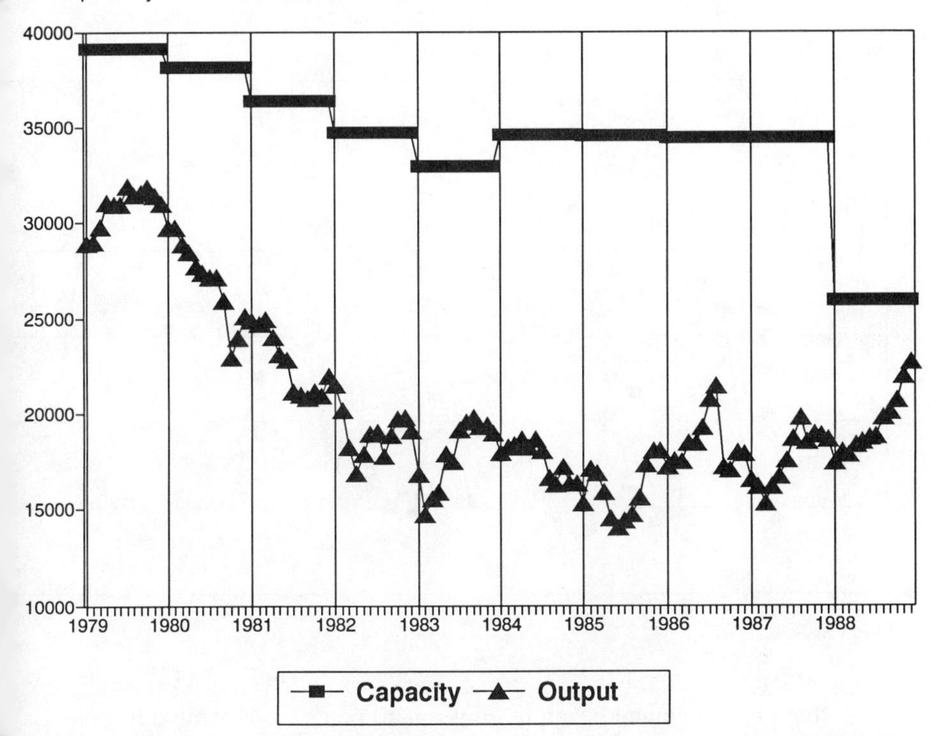

rce of capacity estimates for 1979 and 1988: *OPEC and Lower Oil Prices Impacts on Production Capacity ort Refining, Domestic Demand, and Trade Balances* (Honolulu, Hawaii: Energy Program, East West Center, ruary 1989)

1970s. Some of the automatic responses of consumers were reinforced by government programs ranging from higher excise taxes on petroleum products (especially gasoline) to programs to develop (sometimes seemingly regardless of cost) substitutes for petroleum products. On the supply side, the high prices of the 1970s and early 1980s brought forth an explosive increase in investment in nonpetroleum energy production. By the mid-1980s, worldwide oil demand was caught in a vise of conservation on one side and substitution on the other. Compounding OPEC's problems was that competition for the shrinking oil market intensified as new producers entered the industry. Inexplicably, OPEC made very modest efforts to recruit the new exporters into its market-control programs.

As shown in Figure 3.2, from its peak of 52.3 mb/d in 1979, demand declined by more than 4 mb/d, reaching its lowest point in 1985. At

Figure 3.2. Global oil consumption (in millions of barrels per day).

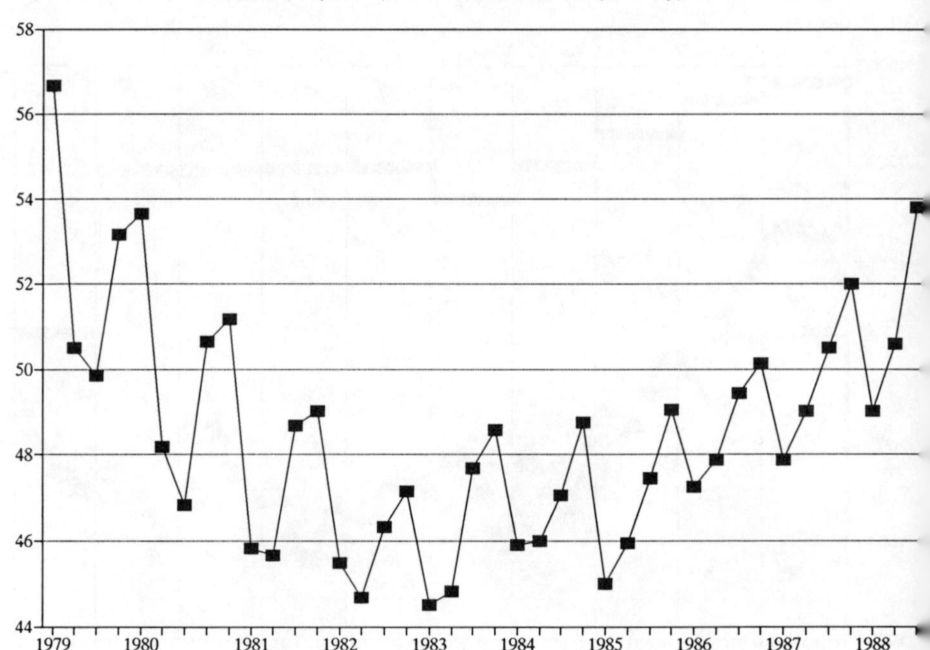

that point, demand began to grow again, reaching 47.9 mb/d in 1986 and 48.8 mb/d in 1987. At the time this was written (September 1989) there was still controversy about the level of growth in oil demand for 1988. A conservative estimate by the International Energy Agency (IEA) put demand at 49.7 mb/d, only 1.8 percent up from the 1987 levels. Other industry groups, however, put the 1988 demand level at 50.6 mb/d, a growth rate of more than 3 percent. While the recovery in oil demand must have been heartening for OPEC, the fact is that it appeared to take a 66 percent drop in prices to promote the growth, and even then it was unclear how robust the recovery was.[6]

Much of the decline in global demand for oil from 1979 to 1985 can be explained by the operation of normal market forces. In an analysis of overt attempts to control the oil market, however, one cannot ignore the explicit steps that governments of oil-importing nations took to diminish oil consumption in their economies. OPEC's price increases of the 1970s and 1980s (whether one looks at OPEC as the master protagonist of these increases or as the clumsy cartel more passively endorsing price increases allowed by market circumstances) created a political force that justified policies to reduce oil dependence.

The 1980s began with a series of oil-conservation programs that had an almost warlike urgency behind them. Indeed, the administration of U.S. president Jimmy Carter called oil import dependence the "moral equivalent of war," justifying massive U.S. investment in synthetic fuels. In Europe, France justified its massive commitment to nuclear energy on the unacceptable risks of oil import dependence. The Japanese government also went the nuclear route, although it would ultimately shy away from using nuclear to the remarkable extent to which the French did.[7] Even among developing countries, massive investments were made to rely less on imported oil. South Korea invested in nuclear power and liquefied natural gas. Brazil, with World Bank assistance, made an enormous and controversial commitment to sugar-based ethanol as a substitute for gasoline. Taiwan also backed away from using residual fuel oil for electricity generation.[8]

If one examines the pattern of worldwide oil demand, one can see that the most obvious effect of national oil-aversion programs was to take the industrial and residential commercial (mostly space-heating) markets away from oil. Oil's share of total fuels consumed by the IEA countries declined from 58 to 54 percent. While this may seem a rather small decline, it marks a revolutionary shift in the energy-economic relationship that had prevailed up to 1973. Prior to that watershed year, oil was the world's growth fuel: It captured the bulk of the increase in energy demand. After 1973, oil became, wherever possible, the fuel of last resort. Only in transportation did it retain its status as the fuel of choice, and there only because of the absence of an alternative.[9] The net effect of these conservation programs was to greatly diminish the prominence of oil in the global energy economy, while the emergence of new oil exporters greatly diminished the prominence of OPEC in the global oil economy.[10]

Compounding OPEC's problems was the refusal of new oil-exporting countries—most prominently Mexico, the People's Republic of China, the United Kingdom, Norway, Oman, Malaysia, Angola, and others—to join the organization. They were not, however, strongly wooed either by the OPEC secretariat (which has never wielded much influence) or by OPEC's leading members. As a result, these new producers were able to proceed as if they could perpetually be outsiders to "OPEC's problem" of maintaining world oil prices. OPEC's share of global oil supplies decreased from 60 percent in 1979 to 38 percent in 1985, the nadir of OPEC's share of the world oil market.

OPEC's share was whittled away by increases in Organization for Economic Cooperation and Development (OECD) production (principally North Sea) and by increases in developing country output (principally in Mexico, Oman, and Angola). The centrally planned economies (CPEs), principally the People's Republic of China, were also able to increase their net exports to the world oil market.[11]

With the sudden decline in oil prices, not only did global oil demand begin to grow again but OPEC's share also began to expand again. When demand for oil turned down, OPEC was indeed the marginal producer in the most negative sense: Its barrels were the first not bought in a competitive world oil market.

By 1985, the broad supply and demand changes set into motion by the enormous oil price increases of 1979 to 1981 had come into play very forcefully. On the supply side, non-OPEC supply had increased from 22 to 28 mb/d, or from 40 percent of total supply to 62 percent. Apparent non-Communist world demand had also decreased sharply, from 52.3 mb/d in 1979 to 46.4 mb/d in 1985. The combined force of reduced demand and increased non-OPEC production caused demand for OPEC oil and condensates to plummet from 32 mb/d in 1979 to 17 mb/d in 1985. This decline in demand for oil was not shared equally by OPEC members. As Figure 3.3 shows, Saudi Arabia absorbed by far the largest portion, and therein lies much of the story about how OPEC's "crisis of '86" began, evolved, and haunts the group still.

As Figure 3.3 shows, from 1980 to 1985 Saudi Arabia was willing to bear the brunt of the decline in demand for OPEC oil. In essence, if OPEC has been the world's marginal supplier of oil, Saudi Arabia has been the swing producer within OPEC. Its willingness to play that role over the years may have conditioned other members, especially the smaller ones, to assume that Riyadh's capacity for doing so was unlimited and that Saudi tolerance for "cheating" by smaller members was limitless. From 1980 to 1985 that did indeed seem to be the case. But by 1985 the level of Saudi exports had reached such a low level that, were the trend to continue, it would have found itself in the absurd position of importing oil to maintain OPEC prices at around $28 per barrel.[12]

Within OPEC, the "free rider" syndrome continued to be a problem because countries "cheating" on their quotas had never had to pay a price for their behavior. In mid-1985, the Saudis resolved to solve both problems simultaneously. Askari (see Chapter 2) has noted that such decisions are made at the highest level in Saudi Arabia; thus Riyadh's decision to abandon official prices was not merely the whim of its famous oil minister, Sheik Zaki Yamani. Askari also

re 3.3. Changes in oil production by selected OPEC members, 1979–1986 (1979 year).

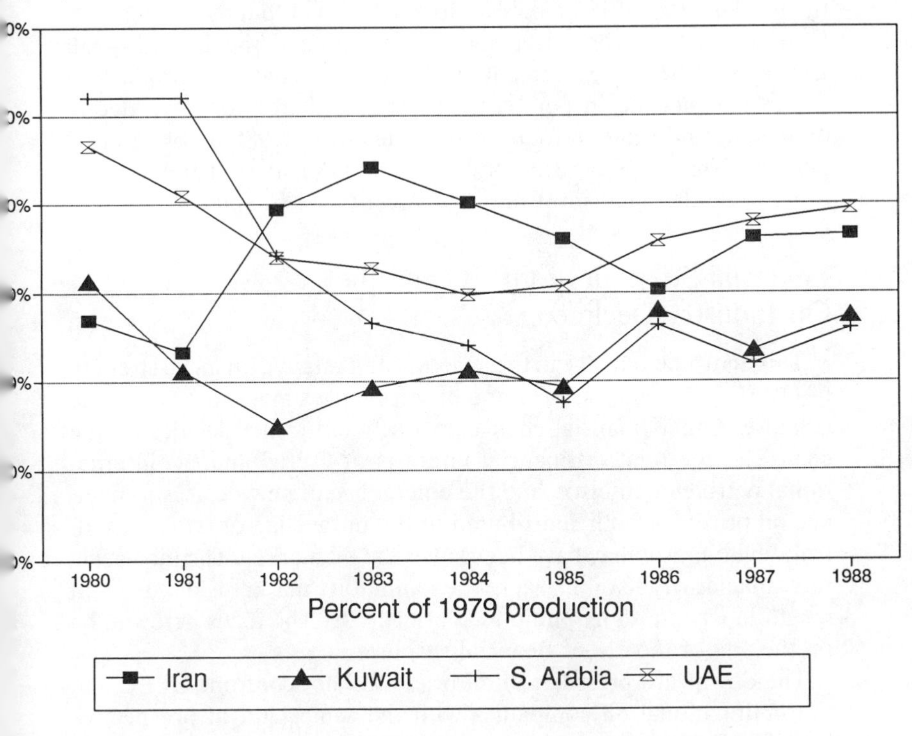

notes that while Saudi oil policy is always motivated by "broad political and economic factors" and while the Saudis are well aware of how to calculate their long-term interests, the Saudis have been unable to generate sources of income other than oil and Saudi oil policy is "therefore influenced by considerations that may maximize short-term revenues as opposed to discounted long-term revenues. . . . Politically, given the importance and size of military and subsidy expenditures, short-run [revenue] needs become at times the overriding consideration." In 1985, Askari argues, the Saudis found themselves faced with short-term revenue needs, and when "external reserves began to decline more rapidly than envisaged under the budget, a general sense of panic emerged." Hence the decision to abandon the role as the virtually sole defender of OPEC's fixed price and quota schemes.

The Saudi market-share strategy put under the harshest possible public light the extreme degree of competition that was emerging within OPEC and among oil exporters in general. As Fadhil J. Al-

Chalabi has noted, OPEC members were not able to increase market shares uniformly. To the contrary, from the third quarter of 1985 to the third quarter of 1986, the Saudis expanded output by 131 percent, Kuwait by 69 percent, Iraq by 50 percent, and the United Arab Emirates (UAE) by 38 percent. The combined change in production shares and decline in per barrel revenues, therefore, saw Algeria experience a decline in total oil revenues from (1985 to 1986) of 60 percent, Nigeria 49 percent, Indonesia 40 percent, and Iran 50 percent, while the Saudis lost only 18 percent (see Chapter 1).

Financial Strength of OPEC and the Oil Industry Declined

This enormous change in the absolute and relative financial strength of OPEC members was only one of the changes induced by the price collapse. Other financial changes contributed to it; specifically, the change in the relative financial interests of OPEC and the international petroleum industry and the emergence of new devices to manage oil price risk both contributed to the unraveling of OPEC's control, tenuous as it may have been, over the oil market. The increasing resemblance of the oil market to a commodity market is discussed at length in Chapter 5 by Philip K. Verleger, Jr. The focus here will be on the end of the era of financial surpluses.

The downward pressure on oil prices did not confront OPEC and the multinational oil companies with the same kind of prospective loss. The independent refiners of the world were better off with lower crude oil prices. The multinational integrated oil companies might regret the loss of upstream margins but they were hedged (to varying degrees) by their downstream operations. The only clear losers from a price collapse were OPEC members, other oil-exporting nations, and the independent crude oil producers, most of whom were in the United States.

It is difficult to prove that this difference in interests had a great operational significance, but one can at least advance the argument that in the 1960s, the multinational companies had a greater stake in the profitability of the upstream segment of their business, *and* they had greater control, and therefore they developed ways to prevent the appearance of the huge overhang of production capacity that emerged in the 1981 to 1985 period. Because they controlled a significant part of the downstream capacity, they could more easily keep downstream demand and upstream supply in balance.

In 1985 OPEC, with virtually no presence downstream, could only resort to price discounting to move its oil into an already glutted refining and marketing network. Once a price war started, large parts of the world oil industry (the independent refiners) could only be happy to fuel further competition for downstream market share.

Once the slide in oil prices began, the reversal in OPEC members' surplus of payments on current account also made it difficult for OPEC members to abstain, either collectively or individually, from the competitive pressures to make more money by selling more and more oil. Once the dam broke, so to speak, there appeared no holding back on the pressure to optimize by volume since it appeared impossible to optimize by price.

The argument, therefore, is that the gradual financial weakening of OPEC members and the decoupling of the financial interests of OPEC and the international oil industry played an important part in the collapse in oil prices. First, it put pressure on all OPEC members to maximize current "cash flow"; second, it removed the international oil industry from the ranks of those who could/would help balance the oil market. Oil companies who used to be part of the solution for market stability had become part of OPEC's problem.

Demise of the Oil Industry's Vertical Integration

The financial "disintegration" between the oil industry and OPEC had a vital operational consequence (for another perspective, see Chapter 9). Oil companies would no longer subsidize upstream operations, even temporarily. If Saudi Arabia wanted to sell oil to its erstwhile Aramco concessionaires, it had to price the oil in such a way that it was entirely competitive with the companies' alternative sources of supply.

The demise of the vertically integrated world oil industry created many nagging operational problems, two of which were most pressing. One was geographic distance. In a market where prices can decline quickly, it matters to an oil company whether it can get crude oil to its refinery in a week, as is the case with oil bought by a European company from Nigeria, or six weeks, as is the case if that refinery buys from a Persian Gulf producer and has to ship oil around the Cape of Good Hope.[13] In the case of short haul cargoes, the buyer is exposed to the risk of a substantial price decline for only a few weeks; in the case of long haul crude, the exposure lasts for several months. All things being equal, when the risk of falling prices is deemed to be high, the buyer will much prefer to buy the short haul cargo.

In the corporate era, the companies owning concessions in Persian Gulf countries did not face such an enormous price risk. Most oil was priced at "official" levels, and thus prices did not change appreciably during the six-week voyage from the Gulf to Europe or America. Even if the price had increased, while the refining affiliate would have suffered, the producing affiliate (e.g., Aramco) would have benefited. In this manner, the integrated structure of the international market of the corporate era ironed out many of the technical pricing issues that bedeviled OPEC in the governmental era.

Second technical factors complicating affairs for OPEC were differences in crude oil quality, two of which are most notable. First, crude oil can have a low specific gravity (i.e., be "heavy") or a high specific gravity (i.e., be "light"). Typically, light crude can be converted into higher fractions of high value "light products," such as gasoline, and thus light crude tends to have a higher market value. The second difference is sulfur content. Oil low in sulfur content ("sweet" crude) is usually more valuable than oil with high sulfur content ("sour" crude).

OPEC has tried for years to fine-tune its schedule of official prices to account for the differences in value between light and heavy, sweet and sour crude. Given the huge volumes of oil sold, however, buyers are very sensitive to small differences in cost. Thus, if OPEC estimates that the spread in price between Arabian Heavy (a sour crude), and Nigerian Bonny (a sweet, light crude) should be $1.50 per barrel but a company figures it can get, say, $2 per barrel more revenue out of Bonny, the company will seek Bonny (and even pay a premium above official prices) and eschew Arabian Heavy.

OPEC's official pricing schedules unavoidably got these technical differences wrong. The values of the different types of crude oils became so fluid in the 1980s that a fixed schedule of prices simply could not keep up with the market realities. As the 1980s wore on, the consequences of "getting it wrong" became more and more serious for the countries whose crude oil was overpriced. Because their oil was not only the farthest from importing areas but also high in sulfur content, the Saudis were particularly hard hit by OPEC's failure to deal with these problems. In essence, the Saudis became the swing producer by necessity, a role they had previously played only by choice.[14] When one country is placed in a position where it more or less unilaterally has to bring global supply into line with global demand, and when that role entails such large cuts in production and consequently revenues that the country swings from having a very large surplus on current account to a very large deficit, then something has to give.[15] In short, the reduction in the call on OPEC oil created by sagging demand and rising non-OPEC supply was concentrated upon Saudi Arabia by the absence of a mechanism whereby OPEC, or an enlarged OPEC, or some other combination of principals, could manage the problem jointly.

Absence of Effective Institutional Mechanism Bringing Supply and Demand into Balance

One of the primary characteristics of what we have called the governmental era (beginning in the early 1970s) is that producers and consumers have had a very imperfect picture of the balance between supply and demand at any point in time. If the world oil market were still vertically integrated, as it was before 1979, information about the balance, or lack of balance, between supply and demand would be automatically transmitted by the refining/marketing branch of the international industry to the producing/transportation segment. This rapid and effective feedback mechanism kept inventory fluctuations to a minimum. If supply exceeded demand, production was trimmed.

When the vertical structure of the industry was decisively ruptured in 1979, this efficient feedback mechanism ceased to exist. As a result, and coupled with important changes in political variables as well as in perceptions of supply availability and price, global inventories increased massively. This stock build would hang over the market like a shroud throughout the first half of the 1980s. It was a major factor in preventing the outbreak of the Iran-Iraq war from raising prices further in 1980. A massive drawdown of excess stocks in the first quarter of 1983 caused the demand for marginal, largely OPEC production to plummet, encouraging the group to lower its official prices for the first time in its history.

The industry learned from this experience that stock changes are both a reflection of changes in supply and demand and a factor that it can look at to anticipate future changes in prices. In the most simple terms, an increase in stocks tends to be taken as a signal that prices are likely to soften, and a decrease in stocks is taken as a signal that prices are likely to become firmer.

By 1985, much of the surplus inventory that the oil industry had accumulated from 1979 to 1981 had been eliminated.[16] After 1985, stocks continued to be a closely watched indicator of the balance between global supply and demand, but the industry had become extremely averse to holding large inventories.[17] As a risk-management device, the ownership of a large inventory is fine if prices are rising but very disastrous when prices are plummeting. Put another way, building stocks was the industry's response to the price uncertainty of 1979, and such use of stocks as a hedging mechanism effectively came to an end several years later, when the industry learned that there was severe down-side risk in holding large inventories. Stocks have never constituted an explicit market-managing mechanism in the oil industry. In theory, the oil market could be kept in balance by a buffer stock, but aside from strategic reserves, which *will* play an important role during governmentally designated emergencies, the importance of stocks will continue to be as signals rather than remedies to imbalances between supply and demand.[18]

This raises the critical question about controls over the oil market in the 1980s. If the oil industry no longer had the structural incentive to cooperate to bring supply into line with demand, and if oil companies and OPEC members were unable and/or unwilling to use stocks for that purpose, the only mechanism available was OPEC's own periodic meetings, which during the course of the 1980s produced one pattern of quota arrangements after another. Even though each and every one of these agreements was aimed at curtailing output, all of them ultimately failed because one or more members was unable or unwilling to abide by the production shares. In essence, the "rules of the game" periodically spelled out at OPEC meetings had a short life.

Rules of the game is a useful concept here because OPEC, like most multilateral organizations, has not just formal agreements but also informal understandings, habits, and an internal pecking order or understanding of which members count more and which count less in the job the organization has been constituted to perform.[19] In the late 1970s and early 1980s, OPEC's essential purpose had shifted from its initial emphasis on gaining control over oil production within its members' borders to determining the price of world oil by setting official price levels for the oils exported by its members. Beginning in the 1980s, when demand for oil turned down, it became necessary to set collective and individual quotas on production in order to sustain the desired price levels.

The most important of such quota-setting exercises in the 1980s occurred in March 1983, when OPEC instituted a quota schedule of 17.5 mb/d. This was, however, only the first of repeated quota-adjusting exercises. It is notable that the first one locked a number of states into production levels far below their capacity. This is particularly true of the UAE, Iraq, Kuwait, and Saudi Arabia. In time, all these countries would chafe at the constraints imposed on them by the quota. As its war with Iran heated up, Iraq simply ignored its 1983 quota and by 1985 was demanding parity with Iran. After much stress and strain, this demand was granted by OPEC in the December 1988 agreement. Another country publicly dissatisfied with the structure set up by the 1983 agreement was the UAE, of whom Chalabi diplomatically observes that "although having committed itself to OPEC's decisions, nevertheless felt that its quota was too low to be observed" (see Chapter 1).

The March 1983 OPEC meeting set the stage for Saudi Arabia to act as the swing producer, a role that it explicitly accepted. Only a year later, however, the continued decline in demand for oil prompted OPEC to call another emergency meeting. The October 1984 agreement produced a revision in production quotas and a reduction in the total OPEC output from the 17.5 mb/d of the March agreement to 16 mb/d. At the October 1984 meeting, some of the issues that would haunt OPEC a short time later began to appear. Iraq was exempted from any further reduction in its quota, although Iran was asked to accept a 100,000 b/d cut. Even non-OPEC members like Mexico and Egypt announced plans to lower production in support of OPEC's efforts.[20] The linchpin of the quota system, however, was still the Saudis' willingness to act as swing producer. The Saudis may have calculated that they would be able to play this role if the demand for OPEC oil would finally begin to rise in 1985. But it was not to be. The amount of oil the Saudis could sell at the prevailing official price continued to decline after the October 1984 emergency meeting. By the third quarter of 1985, Saudi output averaged 2.346 mb/d (see Chapter 1). Askari notes that "this was an intolerable situation for the Saudis: Had they continued producing at this rate, the Kingdom would have been importing oil within two years" (see Chapter 2). By 1985, in other words, the mechanism whereby OPEC sought to control the price of oil in the international oil market, a mechanism that relied on the Saudis to adjust production to bring supply into line with demand, could no longer work for the simple reason that the Saudis would no longer live with it.

Collapse and Recovery and Collapse and Recovery and . . .

By July 1986, the price of some crude had fallen to $8 per barrel. OPEC had increased its output from 14 mb/d in August 1985 to almost 20 mb/d. A bit of this increase (maybe 500,000 b/d) came out of the hides of U.S. producers. Maybe as much as 1 mb/d of OPEC's increase was actually being consumed, but the bulk of OPEC's higher output was going into inventories. In these circumstances, OPEC met again in mid-July 1986.

At this meeting, Iran surprisingly agreed to exempt Iraq from having to adhere to its quota. With Iraq, whose production was running at nearly double its authorized quota, no longer in the picture, the other OPEC members agreed to maintain production at the levels agreed to in October 1984. This was sufficient to raise oil prices back to $15 per barrel. In October 1986 OPEC met again, struggled through various crises, and in the end "rolled over" the August accord until December. In December the OPEC meeting yielded a new accord that included further production cuts, the elimination of netback deals, and a commitment to a fixed price of about $18 per barrel. Iraq was still exempt but the other members were on board and some non-OPEC producers signaled cooperation.

The December agreement did not decisively resolve the swing producer issue. In the next few months, Saudi Arabia wound up curtailing its production to levels far below quota, while virtually every other OPEC member (except Iran, whose output was curtailed more by war-related constraints than by deliberate policy) produced at or above quota. Crude oil prices, which had rallied at the end of 1986 from $11 per barrel (for Arabian Light FOB the Persian Gulf) to $15 per barrel after the OPEC meeting, climbed to over $16 per barrel in January and remained above $16 through the first quarter, when demand for crude oil is typically the lowest of the year.[21]

International oil prices remained firm through the second and third quarters of 1987, partly on the strength of market anxieties about the Iran-Iraq war. In the spring of 1987, the United States decided to dispatch warships to the Persian Gulf to escort Kuwaiti oil tankers, which were being subjected to increasingly effective attacks from Iran. This provoked a number of attacks on U.S. warships and on oil tankers in the summer of 1987. The apparent increase in U.S.-Iranian tensions lent further strength to the price of oil.

In spite of this price strength, the extent and size of quota violations by OPEC members increased. By the end of July, Iranian production was 140,000 b/d above quota, Kuwaiti output 400,000 b/d, UAE 400,000 b/d, Venezuela 230,000 b/d, and Indonesia 140,000 b/d. OPEC's total output in July reached 18.8 mb/d, a full 2.2 mb/d above its self-imposed quota of 16.6 mb/d.[22]

By the end of 1987 OPEC production was once again well in excess of what the market would require. The most egregious cheating was by the UAE, which was producing at nearly double its OPEC-approved rate. Prices began to fall from the $18 per barrel level that OPEC had officially decreed to be the official oil price. By the time OPEC gathered for its usual December meeting, prices had fallen to $14 per barrel, and since the market was then heading into its usual seasonal decline in demand for crude oil (when requirements can fall by as much as 3 mb/d), the ministers met again under the cloud of an immanent price collapse.

At the December 1987 OPEC meeting, the members rolled over

the production and quota schedules agreed to at the end of 1986. That is, OPEC output excluding Iraq would, in theory, be held at about 15.1 mb/d. Iraq, it was presumed, would produce at least 2.5 mb/d, bringing the OPEC total to at least 17.6 mb/d. Adding in Neutral Zone production would bring OPEC production, under the best of circumstances, to about 18 mb/d in the first quarter of 1988.

Initially, the market's reaction to the roll over announcement was positive. Prices, which had fallen from $17 per barrel (Arabian Light, FOB the Persian Gulf) to $14 per barrel in December, rallied to the $15 range, where they stayed through January. The trade press particularly played up the UAE's apparent willingness to abide by the agreement.[23] But the confidence was short-lived. By mid-February, prices were falling again.

By the end of March 1988, the spot price of Dubai oil, by this time widely used as an international benchmark, had fallen from the $17 it reached at the end of December to $13. This decline occurred in spite of the fact that most OPEC members did abide by their production quotas. The market, however, wanted less than 18 mb/d, as became evident from the sharp drop in oil prices in February and March 1988.

In April 1988 there was a meeting between OPEC and selected non-OPEC oil exporters (Mexico, Oman, Angola, Malaysia, and China attended, and even the Texas Railroad Commission sent an unauthorized representative). For some weeks there was intense speculation in the market (reflected in the strengthening of oil prices in March and April) that oil's old guard and the newcomers would strike a deal. In the end, however, the non-OPEC countries mustered only an offer to cut exports by 5 percent. This amounts to some 200,000 b/d, a volume that Saudi oil minister Hisham Nazir diplomatically welcomed but which, for a country already restraining production by 4 mb/d, could not have been very impressive.

In the summer of 1988, Abu Dhabi, Saudi Arabia, and Kuwait began substantially to increase their production to levels far in excess of their OPEC quotas. There is still disagreement about the motives for their behavior. Perhaps they wanted to send a signal to Iran and Iraq that the end of the war would not provide unlimited opportunities to regain market share at the expense of the other Arab Gulf states. Another argument is that the quota violations of Saudi Arabia and Kuwait were motivated by their desire to discipline Abu Dhabi, whose "go it alone" strategy had been a constant source of intra-OPEC tension in the 1980s. Yet another argument is that Saudi Arabia was simply in need of revenues and had become so disenchanted and cynical about OPEC that it scarcely paid attention to its quota.

Whatever the motivations, the effects of these increases in OPEC output was to cause prices to fall to the lowest levels seen since 1986. The price of Dubai oil fell below $10 per barrel in the autumn of 1988. As prices fell, there was great concern in the market that OPEC would lose its grip on the market altogether. The greatest concern, as OPEC's end-of-the-year meeting neared, was that Iraq would not be reabsorbed into the quota system and that its pursuit of its own interest would trigger a general scramble for market share and a disastrous fall in oil prices. On its part, Iraq had steadfastly maintained that it would rejoin OPEC only if it was given production parity with Iran. In the original quota allocations of 1983, of course, Iraq's allocation was only half of Iran's.

To those who believed Iran would have difficulty agreeing to give Iraq production parity, the November 1988 OPEC meeting appeared destined for failure, regardless of the surprising strength of world oil demand. But in a surprising move, Iran agreed to a two-prong proposal to give Iran and Iraq production parity and raise the overall OPEC quota to 18.5 mb/d.

The November 1988 agreement provided something old and something new. What was old was the apparent unwillingness of OPEC members to address directly the grievances of perpetually unhappy Abu Dhabi and the United Arab Emirates. While the UAE was granted the same small increase in quota, and while the Abu Dhabi delegate to the OPEC meeting appeared to agree to the revisions, it soon became apparent that UAE production would *not* be reduced to quota levels.

The UAE problem could be seen as an example of a mismatch between the quotas and production capacities of Saudi Arabia, Kuwait, and Qatar as well. All these countries had quotas that were about 50 percent of their production capacities. So while the November 1988 agreement took care of one problem, bringing Iraq back into the OPEC fold, it did not take care of the old problem of Abu Dhabi, nor did it raise the quotas of the other Arab countries who could, on the basis of quota to production capacity ratios, argue that they were carrying a bigger share of OPEC's market-managing load than they should.

Thus, at the beginning of 1989 one could not say that OPEC had firmly recaptured control over the market. On the basis of OPEC's difficulties in sticking to the 1986 and 1987 agreements, one had to characterize the end-1988 agreement as a rather fragile one. The indications that demand was rising suggested, but only suggested, that OPEC's worst days were behind it. But at the same time, Iran and Iraq were likely to boost output significantly in the 1990s, and other Arab Gulf countries were capable of doing the same; thus production capacity was likely to continue to exceed even rising demand. Under these circumstances, could OPEC regain in the 1990s the kind of control it had over the market in the heyday of the corporate era?

Conclusions

This review of the ups and downs of the oil market in the 1980s began with a note on the levels of analysis problem. International oil is a sufficiently complicated system to have elicited a number of competing visions of what drives the system, how it works.

One of the most fundamental questions about oil is about the consequences of its inherent economic character; that is, the particularly capital-intensive nature of the oil business. There are those who believe that the high ratio of fixed to variable costs has, in Paul Frankel's words, "all-pervading repercussions":

> The extreme relation of fixed and variable costs . . . in inverse ways determines the operators' actions and reactions: on the one hand the rapidly falling cost curve . . . encourages intensely competitive behavior; on the other, and for that selfsame reason, there is a deepseated tendency to avoid, or at least to mitigate, the results of this situation not only by horizontal concentration and vertical integration but also by cooperative endeavors leading to various degrees of "understandings" among the competitors, some of them engendered by government authorities.[24]

Put another way, Frankel argues that precisely because the oil industry tends toward such intense competition that in the end only one monopolist remains, it tends to generate adjustment measures like cartelization and regulation. Morris Adelman argues to the contrary: Because the industry tends toward such intense competition that in the end many competitors remain, adjustment measures are not only unnecessary but economically inefficient.[25] This most fundamental question addresses the character of the *industry*, while the analysis presented here concerned itself with the oil *market*. What little light it sheds on this fundamental debate is that Adelman may be right in the long run, but Frankel's view is more revealing in the short run.

The record of the 1980s shows the immense leeway that Saudi Arabia and a few other producers had in pursuing what turned out to be a very costly campaign to defend high prices in the early 1980s. Ultimately, as Adelman's view of the industry would suggest, the Saudis were forced by competitive pressures to back off this campaign. Thus the 1986 collapse in prices would enable Adelman to say that his view of the industry—that it is inherently competitive whether government hands are on the helm or not—is correct. But the reemergence of OPEC in 1986, its failure in 1987, its success in getting back on track in 1988, the very act of failing and trying again to reorganize what Frankel calls "adjustment measures" indicate his view of the industry is correct. We see again and again new efforts to organize the markets. We go from something short of a cartel to something short of anarchy and back again.

The second level of analysis of the oil market is whether supply and demand conditions in the petroleum market are such that the opportunity to create a cartel can exist. At this level, the issues are the elasticities of petroleum demand and supply to price. There are fewer doctrinal trench lines here. Economists seldom lose faith that, sooner or later, elasticities of some sort do exist and that they will assert themselves sooner or later. The debate is between how much sooner and how much later. In this area, John Lichtblau (Chapter 10) provides a self-admitted "consensus" forecast showing "a modest increase in world oil demand—1 percent per year or slightly more." In view of the fairly low oil prices behind this forecast ($22.40 in 1995 in constant 1987 dollars), Lichtblau may be said to believe in a very low elasticity of demand to price. This view may be contrasted with that of William Hogan, who has argued that elasticities may be substantially higher than the consensus view and that demand for oil may be considerably above the 1 percent growth specified by Lichtblau.[26]

The analysis presented here provides preliminary evidence that Hogan's view may turn out to be closer to the truth. Oil demand in 1988 increased by substantially more than the 1 percent of the consensus forecast. As of early 1989, estimates of 1988 demand growth ranged from 2.9 percent to 3.8 percent.[27] This rather high growth rate is certainly rooted in part in the higher-than-expected world economic growth in 1988. It is also possible, however, that elasticity of demand to the lower oil prices since 1986 is causing a fundamental change back to a higher oil consumption growth rate.

The third level of analysis of the oil market deals with OPEC behavior. If supply and demand conditions provide opportunities to organize the industry as a cartel, is the effort to organize the cartel effective or not? A great deal of OPEC theorizing has been focused at this level around the question of discount rates. Adelman, Griffin and Teece, Pindyck, Eckbo, and others have theorized about the implications of different groups of cartel members having (or behaving as if they have) different discount rates.[28] Much of this analysis divides OPEC into "dominant producer" and the rest, spenders and savers, the cartel core and the rest, price hawks and price doves, and so forth. Much of this was written at a time when the price increases of 1979–80 had indeed given the low population, high oil production countries revenues in excess of their immediate ability to spend. By the middle of the decade, however, such distinctions appeared less clear-cut. The Middle Eastern producers showed a prodigious talent for eliminating their surpluses, with the practical effect that as of 1986 one could say that all OPEC countries have a high discount rate, they are all spenders, and none are "core members" of the cartel, in the sense of being able or willing to discount present revenues for future ones.

These conclusions suggest that the 1990s present an even more complex international oil market than the 1980s. If demand and supply do turn out to be more elastic to price than the conventional view suggests, the demand for OPEC oil should increase more quickly than expected. If the Frankel view of the industry is right, the industry should continue to generate adjustment mechanisms that would tend to diminish competition. In the 1970s, the "industry" meant oil companies somewhat antagonistic to yet integrated with oil-exporting countries; in the 1980s the industry meant oil companies dealing with oil-exporting countries on an arm's-length basis. The industry in the 1990s will increasingly mean oil companies associated with oil-exporting countries. Vertical reintegration by itself, however, could mean more and more competition throughout all facets of the integrated industry, from wellhead to service station. The industry (meaning countries and governments) will probably try to generate more effective adjustment mechanisms in the 1990s. This will not be easy. If no one has the financial strength to discount the present for the future, if everyone has a high discount rate, then integration by itself is unlikely to be a sufficiently powerful mechanism to regulate the market. That logic leads to the primary conclusion of this analysis: The international oil market of the 1990s will see again and again new efforts to organize the markets, but these efforts will swing between something short of an organized cartel and something short of anarchy and back again.

This outlook is based on several arguments. First, if demand for oil grows fast enough, and it may, then the amount of surplus production capacity will diminish and be increasingly concentrated in what used to be called "strong hands" (i.e., financially strong countries like Kuwait). But the decrease in excess capacity by itself may not be enough. With financial mismanagement by Saudi Arabia, Iran, Iraq, and the UAE, there might not be any strong hands to curb production when necessary.

Even if there are strong hands in the crude end of the business, it is also possible that increases in the pace and extent of downstream integration by OPEC members could make them increasingly indifferent to crude oil prices and more interested in refining and marketing margins. Downstream integration by all OPEC members without "mechanisms" like the major oil companies used to have in the 1960s may mean more competition, not less, along the entire spectrum of activities in the oil business. But how can OPEC's oil companies, all government-owned, create the kinds of contractual mechanisms that make it possible, in Moran's words, "to limit the scope for independent action when collusive agreements come under stress, to make credible commitments in the face of future uncertainties, to bind themselves to the joint objective"?[29]

It seems unlikely that all four of these conditions, all of which in our argument are needed at the same time to give anyone effective control over oil prices, will be satisfied in the 1990s. Thus, the prospects for the 1990s are for continued instability of international oil prices.

Notes

1. Paul Frankel, *Essentials of Petroleum* (London: Frank Cass, 1946).
2. John Blair, *The Control of Oil* (New York: Pantheon Books, 1976).
3. James M. Griffin and David J. Teece, Eds., *OPEC Behavior and World Oil Prices* (London: George Allen and Unwin, 1982); and Ian Skeet, *OPEC: Twenty-five Years of Prices and Politics* (Cambridge: Cambridge University Press, 1989). Skeet's book is contemporary historiography, if that is not an oxymoron, because it evaluates OPEC's twists and turns right up to the late 1980s.
4. Charles Doran, *Myth, Oil, and Politics: Introduction to the Political Economy of Petroleum* (New York: Free Press, 1977).
5. There are few estimates of production capacity. One estimate that has the virtue of having been published throughout the 1980s is that of the U.S. Central Intelligence Agency, published in the agency's monthly *International Energy Statistical Review*. That publication has provided three different estimates of OPEC production capacity. The first is "installed" capacity, which includes all aspects of crude oil production, processing, capacity, and storage. This is generally the highest capacity estimate. The second is "maximum sustainable" capacity, which is the maximum production rate that can be sustained for several months. It considers the experience of operating the total system and is generally some 90 to 95 percent of installed capacity. This capacity concept does not necessarily reflect the maximum production rate sustainable without damage to the fields. "Available" capacity reflects production ceilings announced by governments. Iraqi and Iranian capacity

throughout this period was limited by damage to facilities done by the Iran-Iraq war.

6. There is widespread controversy about whether the strong 1988 demand growth was due to temporary factors like unusually strong global economic growth (suggesting oil demand would taper off once the global economy cooled down) or whether the growth marked a more ingrained return to oil in favor of other energy sources. The U.S. Department of Energy, for example, had been arguing that a 3–2–1 ratio existed in the U.S. energy economy, in which a 3 percent economic growth required a 2 percent energy growth rate and a 1 percent oil growth rate. On the basis of the 1988 experience, it appears that ratio will be closer to 5–2–3.

7. The most comprehensive review of the energy policy conservation programs of industrial countries can be found in International Energy Agency (IEA), *Energy Policies and Programmes of IEA Countries* (Paris: OECD, 1988), esp. pp. 30–87.

8. In many developing countries, the key to improving energy efficiency was often simple price reform rather than elaborate conservation policies. For a review, see the *Energy Journal* 9 (1988).

9. There are only two alternatives to oil in private automobiles. One is natural gas, but with the exception of New Zealand, where gas exists in such abundance *and* with no other economic outlets, no other country has developed a large-scale gas-powered car fleet. The other alternative is alcohol, which has gained a prominent place in Brazil's energy economy but at a price that few other countries have been willing to pay.

10. See International Energy Agency, *Energy Policies*, p. 58.

11. Ibid.

12. I am indebted to Adam Sieminski for putting the Saudi position in such a stark manner.

13. About half of Saudi Arabia's exports and all of the exports of Iran, Kuwait, and the United Arab Emirates (UAE) have to be shipped by tanker through the Straits of Hormuz, and in the case of European and American buyers, around South Africa. The voyage takes about six weeks.

14. In fact, during the 1979–80 market turmoil it was deemed to be an advantage to be able to buy oil from Saudi Arabia because the Saudis were the last to raise prices in a rising market. When they also became the last to lower prices in a falling market, however, buying oil from Saudi Arabia became a costly proposition.

15. The Saudi current account position swung from a $40 billion surplus in 1981 to a $10 billion surplus in 1982 to a $20 billion deficit in 1983. It has remained in deficit since then. International Monetary Fund, *International Financial Statistics*, various issues.

16. It is a matter of political and strategic interest that governments built up their strategic stocks by about as much as industry drew down its "discretionary" stocks. The growth of strategic reserves is a story that would have greater significance for 1987, when the war in the Persian Gulf intensified and U.S. involvement in the war increased. In the events leading to

the price collapse of 1985–86, however, strategic stocks played virtually no part.

17. Worldwide stocks are reported by the IEA's *Oil Market Report*, a monthly publication, and by Energy Security Analysis, Inc.'s *Stockwatch* and *Euroilstock* services.

18. For a more theoretical discussion of buffer stocks, see Charles P. Kindleberger and Peter H. Lindert, *International Economics*, 6th ed. (Homewood, Ill.: Richard D. Irwin, 1978), esp. Appendix F.

19. The concept is very effectively applied by Kenneth Dam in *The Rules of the Game: Reform and Evolution in the International Monetary System* (Chicago: University of Chicago Press, 1982).

20. "OPEC Cuts Production Quotas in Bid to Hold Prices," *Petroleum Economist* (December 1984): 444.

21. Although the demand for finished products is usually highest in the winter due to high demand for heating oil, demand for crude oil is highest in the autumn because refiners must order their supplies from long haul producers like Saudi Arabia well before the peak seasonal demand for refined products.

22. Production figures from *Petroleum Intelligence Weekly*, various issues.

23. For example, an article headlined "OPEC Policy Shifting from Fixed Prices to Volume Restraint," *Petroleum Intelligence Weekly*, 25 January 1988, p. 2 noted that "Abu Dhabi and Kuwait are losing [sales] volume by refusing to fully accommodate buyers' pricing ideas."

24. Paul Frankel, "The Oil Industry and Professor Adelman: A Personal View," in Ian Skeet, ed., *Paul Frankel, Common Carrier of Common Sense* (Oxford: Oxford University Press for the Oxford Institute for Energy Studies, 1989), p. 214.

25. Morris Adelman, *The World Petroleum Market* (Baltimore: Johns Hopkins University Press, 1972).

26. William Hogan, "Oil's Demand and OPEC's Recovery," Harvard University, John F. Kennedy School of Government, Energy and Environmental Policy Center, Discussion Paper Series E-88-2, June 1988.

27. See International Energy Agency, *Oil Market Report* (January 1989).

28. For an excellent set of arguments about OPEC, see James M. Griffin and David J. Teece, *OPEC Behavior and World Oil Prices* (London: George Allen and Unwin, 1982). Among the recommended essays are the authors' introduction and Morris Adelman's "OPEC as a Cartel."

29. Theodore H. Moran, "Managing an Oligopoly of Would Be Sovereigns: The Dynamics of Joint Control and Self-Control in the International Oil Industry," *International Organization* 41 (1987):575–607.

DAVID H. VANCE

4. Long-Run Oil Prices and U.S. Energy Security

Oil Price Perspective

WITHIN the last decade and a half, the oil market has undergone three major upheavals: the energy crises of 1973–74 and 1979–80, with the rapid price runups these involved, and the price collapse of 1986. Was the collapse really a new upheaval, or was it merely the inevitable correction of earlier excesses? Figure 4.1 displays the path of real prices over the century. Table 4.1 lists oil prices from 1901 through 1986 in both current (nominal) and constant (real) dollars. The average real price of oil in 1985 dollars throughout the entire 1901 to 1986 period was $14.99. The average for 1901 to 1972, before OPEC gained control of oil pricing, was $12.56. The average price for 1973 to 1986, with OPEC in control, was $27.50. This long-run price history indicates that the high prices of the late 1970s and early 1980s were an aberration and that the price collapse of 1986 was a correction to a more normal level.

These price data alone cannot answer the question of whether or not OPEC will be able to maintain prices significantly above its pre-OPEC average in the long run. But they seem to indicate that OPEC's reach exceeded its grasp in the 1970s, and they raise the possibility that oil prices are likely to remain below their recent OPEC highs for many years.

The price collapse notably did not even approach historic lows in oil prices. The 1986 real price was still some 30 percent above that of 1972, the last year before OPEC took control. Furthermore, OPEC has been able to raise and hold prices significantly above the 1986 level this year.

Finally, this history is based on U.S. prices and thus does not reflect the non-U.S. world price during the 1959 to 1973 period when the Mandatory Oil Import Control Program (MOIP) kept U.S. prices at least $1.30 (nominal) above world prices. Although the MOIP did protect the U.S. industry for a time, its effect on U.S. energy security is questionable. It resulted in a more rapid depletion of U.S. limited reserves, so that when the crunch did come in 1973 the industry was less able to respond from U.S. resources. Furthermore, by keeping U.S. demand for imports down, the MOIP helped hold world oil

Figure 4.1. Real price of oil, 1901–1986 (in 1985 dollars).

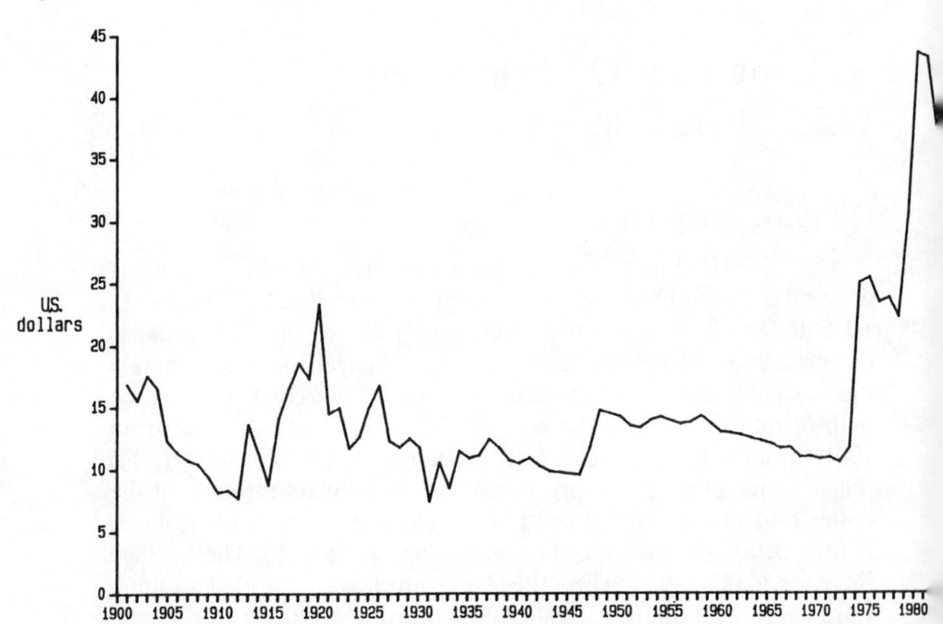

Source: Table 4.1.

prices lower than they might otherwise have been, resulting in greater reliance on oil in Europe and Japan. In this sense, it greatly increased the abruptness with which the first energy crisis in 1973 hit the world economy.

How Did the Price Jumps Happen?

Extremely rapid growth in oil demand from 1960 to 1973, especially outside the United States, set the stage for the price jumps. World oil consumption was growing some 7.7 percent per year during the period. From 1960 to 1973, oil's share of the growth in energy consumption was 84.7 percent in other Organization of Economic Cooperation and Development (OECD) countries versus 49.6 percent in the United States. In 1960 these countries were only 35.1 percent dependent on oil versus a 42.8 percent dependency for the United States. By 1973 they had become 62.1 percent dependent versus 45.6 percent for the United States.

The rapid growth in oil demand (far outstripping the growth in oil reserves) and rising dependence on oil (especially outside the United

Table 4.1. U.S. Prices, Nominal and Real, 1901–1986

Year	*Nominal $*	*1985 $*	*Year*	*Nominal $*	*1985 $*	*Year*	*Nominal $*	*1985 $*
1901	1.14	16.86	1930	1.62	11.75	1959	4.00	13.66
1902	1.09	15.53	1931	0.93	7.43	1960	3.87	12.99
1903	1.24	17.57	1932	1.17	10.53	1961	3.87	12.90
1904	1.18	16.51	1933	0.92	8.46	1962	3.89	12.73
1905	0.90	12.27	1934	1.36	11.49	1963	3.88	12.49
1906	0.85	11.32	1935	1.31	10.89	1964	3.87	12.28
1907	0.84	10.72	1936	1.45	11.15	1965	3.89	12.05
1908	0.81	10.42	1937	1.58	12.45	1966	3.88	11.67
1909	0.76	9.45	1938	1.46	11.74	1967	3.90	11.78
1910	0.67	8.12	1939	1.32	10.73	1968	3.93	10.99
1911	0.68	8.32	1940	1.32	10.48	1969	4.15	11.04
1912	0.65	7.64	1941	1.48	10.94	1970	4.29	10.83
1913	1.15	13.58	1942	1.53	10.27	1971	4.56	10.96
1914	0.98	11.34	1943	1.54	9.83	1972	4.57	10.56
1915	0.79	8.75	1944	1.56	9.73	1973	5.38	11.74
1916	1.42	14.03	1945	1.58	9.62	1974	12.52	25.17
1917	2.07	16.48	1946	1.82	9.54	1975	13.93	25.61
1918	2.64	18.56	1947	2.50	11.66	1976	14.48	23.55
1919	2.79	17.31	1948	3.39	14.77	1977	14.53	23.96
1920	4.30	23.40	1949	3.30	14.51	1978	14.57	22.38
1921	2.21	14.45	1950	3.31	14.24	1979	21.67	30.63
1922	2.10	14.93	1951	3.34	13.51	1980	33.89	43.84
1923	1.68	11.66	1952	3.34	13.32	1981	36.80	43.43
1924	1.81	12.59	1953	3.55	13.95	1982	33.59	37.03
1925	2.19	15.02	1954	3.68	14.24	1983	29.35	31.46
1926	2.41	16.78	1955	3.67	13.91	1984	28.87	29.82
1927	1.72	12.25	1956	3.71	13.63	1985	27.04	27.04
1928	1.68	11.76	1957	3.89	13.80	1986	14.67	14.27
1929	1.78	12.49	1958	4.09	14.31			

Source: 1901–1981—Based on data in Arlon R. Tussing, "Reflections on the End of the OPEC Era," *Alaska Review of Social and Economic Conditions*, Dec. 1982. 1982–1986; Oil Prices Are Refiner Acquisition Cost of Crude Oil from DOE/EIA, *Weekly Petroleum Status Report*, GNP Deflators from Council of Economic Advisors, *Economic Indicators*, (monthly).

States) set the stage, but the proximate causes of the 1974 and 1979–80 price jumps were political/military crises that threatened oil supply—the Arab oil embargo during the Arab-Israeli war and the revolution in Iran. *Both price jumps were completely out of proportion to any supply decline*. Figures 4.2 and 4.3 display aggregate consumption and production for these periods (data are provided in Tables 4.2 and 4.3). At this broad level of aggregation it is difficult to see any significant drops in either consumption or production during the "shortage" periods, neither of which shows declines outside the range of normal seasonal variations. The gas lines—the most

Figure 4.2. World oil production and inland oil consumption among major Organization for Economic Cooperation and Development (OECD) countries, namely: United States, Canada, Japan, France, Italy, United Kingdom, and West Germany (some 79 percent of OECD total), 1973–1974. *Abbreviation:* b/d, barrels per day.

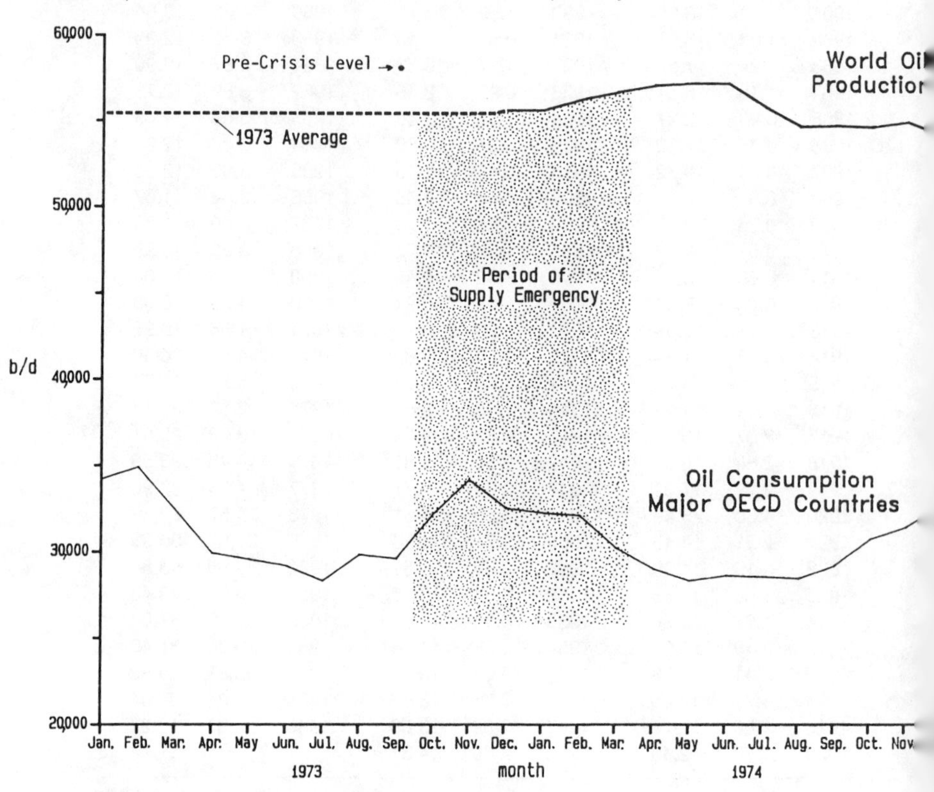

Source: International Oil Developments, CIA.

public manifestation of the "shortage"—were a function of misguided effort at controls rather than any lack of oil, as the Department of Energy (DOE) security report makes clear.[1]

What OPEC Did and Didn't Do

A highly plausible case can be made that buyers of oil initially were more responsible for the price increases than OPEC because spot prices rose ahead of contract prices, which were adjusted to follow them with a significant lag. Part of the explanation for this lies in industry inventory behavior—in both cases the end of the crisis period found inventories much higher than at the beginning, which means ıre 4.3. World oil production and oil consumption among selected OECD countries, ıely United States, Canada, Japan, Austria, Belgium/Luxembourg, Denmark, France, , the Netherlands, Norway, Spain, Sweden, the United States, West Germany, and tralia (some 90 percent of OECD total), 1978–1979.

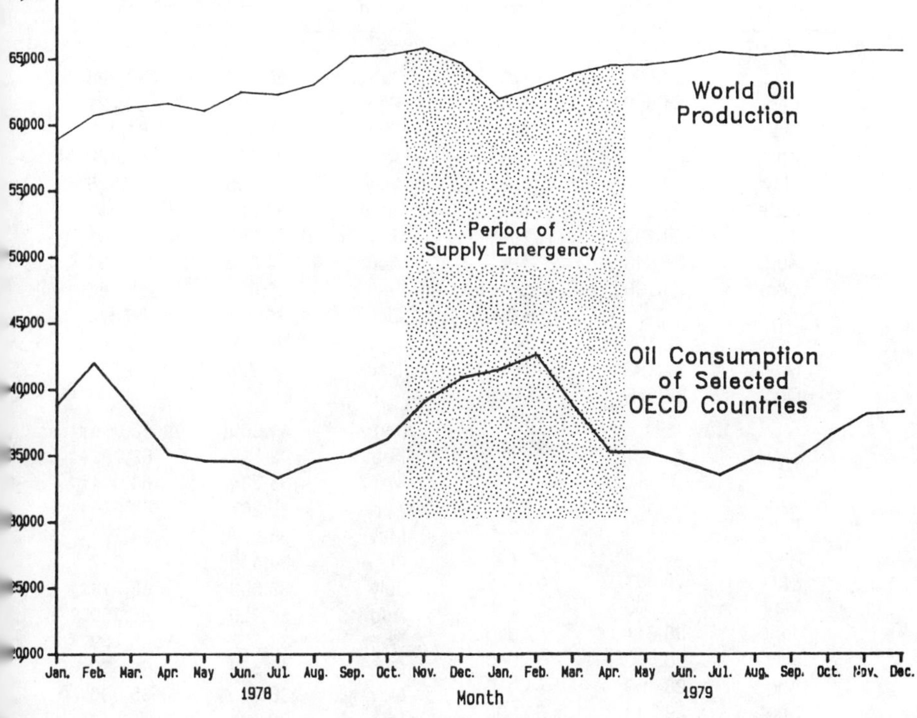

es: *Petroleum Economist: International Energy Statistical Review*, CIA.

that stockpiling demand was a major factor in pushing up prices. Part of the explanation lies also in the differential impact across companies and countries, especially in the case of the Iranian revolution. Those companies or countries that were particularly dependent upon Iran did experience real shortages at least in the beginning and had to scramble for new supplies, while those less affected tended to hang onto their supplies because of the crisis atmosphere and expectations of higher prices. In each case, however, OPEC clearly attempted to hold prices up to the crisis-induced level after the crisis had passed and in the face of falling demand, reducing production to do so. *It is this effort to hold the price at a much higher level that distinguishes the OPEC period from those that preceded it.* Furthermore, these crises must be set against the background of the long-run projections

Table 4.2. Oil Production and Consumption, 1973–1974 (thousands of barrels per day)

Year and Month	Oil Consumption, Major OECD Countries*	World Oil Production
1973		
Jan.	34,162	NA
Feb.	34,880	NA
Mar.	32,386	NA
Apr.	29,906	NA
May	29,545	NA
June	29,175	NA
July	28,293	NA
Aug.	29,843	NA
Sept.	29,596	(Precrisis) 58,080
Oct.	32,198	NA†
Nov.	34,192	NA†
Dec.	32,838	(Avg. 1973 = 55,685)†
1974		
Jan.	32,251	55,711†
Feb.	32,125	56,330†
Mar.	30,253	56,780†
Apr.	29,051	57,180
May	28,330	57,330
June	28,611	57,310
July	28,580	55,940
Aug.	28,480	54,810
Sept.	29,243	54,900
Oct.	30,791	54,805
Nov.	31,496	55,135
Dec.	32,798	54,305

Source: International Oil Developments, CIA.

* Inland oil consumption for United States, Canada, Japan, France, Italy, United Kingdom, and West Germany (some 79 percent of Organization of Economic Cooperation and Development [OECD] total).

† Period of supply emergency per U.S. Department of Energy energy security reports.

Table 4.3. Oil Production and Consumption, 1978–1979 (in thousands of barrels per day)

Year and Month	Oil Consumption Selected OECD Countries*	World Oil Production
1978		
Jan.	38,837	58,883
Feb.	42,011	60,721
Mar.	38,655	61,353
Apr.	35,099	61,670
May	34,595	61,068
June	34,558	62,531
July	33,293	62,301
Aug.	34,569	63,131
Sept.	35,053	65,266
Oct.	36,271	65,374
Nov.	39,130	65,940†
Dec.	40,929	64,739†
1979		
Jan.	41,565	62,054†
Feb.	42,712	62,992†
Mar.	38,784	64,004†
Apr.	35,267	64,686†
May	35,294	64,727
June	34,496	65,097
July	33,559	65,713
Aug.	34,920	65,450
Sept.	34,527	65,769
Oct.	36,558	65,582
Nov.	38,240	65,936
Dec.	38,434	65,941

Sources: Petroleum Economist; *International Energy Statistical Review*, CIA.

* Inland oil consumption for United States, Canada, Japan, Austria, Belgium/Luxembourg, Denmark, France, Italy, Netherlands, Norway, Spain, Sweden, United Kingdom, West Germany, and Australia (some 90 percent of Organization of Economic Cooperation and Development [OECD] total).

† Period of supply emergency per U.S. Department of Energy security report.

and expectations of the time, which were themselves largely shaped by OPEC. Much has been made of the amount of shut-in OPEC capacity. In fact, many studies appear to consider the percent utilization of existing OPEC capacity as the single most important explanatory variable in predicting oil price behavior. But what usually gets overlooked is the difference between existing capacity and potential capacity. The amount of capacity that OPEC could easily have developed (but didn't) is double the amount of installed capacity it shut in (see below).

What Finally Upset OPEC's Control

When oil use dropped and non-OPEC oil production rose in response to the higher oil prices, OPEC lost market share. Cheating by some OPEC members further squeezed the share of Saudi Arabia, then the swing producer, until the Saudis finally abandoned the swing role, adopted netback pricing, and brought about a collapse in oil prices.

Demand Side Effects. Both OPEC and many forecasters overlooked the macroeconomic impacts of the oil price explosions. The large price increases lowered real incomes in consuming countries, which in turn lowered oil demand by shifting the demand curve sharply to the left (i.e., lowering the quantity demanded at any given price). A large part of the reduction in the demand for oil was caused by lower economic growth, much of which was a direct result of the oil price increases. Estimated world GNP grew some 4.9 percent per year from 1960 to 1975, but only 2.8 percent per year from 1975 to 1984. Many factors affect world growth and not all of the decline can be blamed on oil prices, but clearly they were a major factor.

The long-run demand curve for oil also turned out to be much more elastic than the short-run curve, attributable in large part to conservation and fuel substitution. For OECD countries (the major industrialized countries), the ratio of total primary energy consumption (TPE) to gross domestic product declined 25.1 percent from 1973 to 1984 after having risen 1.4 percent from 1960 to 1973. The ratio of oil consumption to GDP fell even more dramatically—by 56.8 percent from 1973 to 1984 after having risen by 27.9 percent from 1960 to 1973. The decline in the TPE/GDP ratio largely reflects the effects of conservation, while the faster drop of the oil/GDP ratio shows the added effect of fuel substitution.

Oil is used both in stationary plants (electric generators, industrial uses, and residential/commercial heating), where substitutes are readily available, and in transportation (cars, trucks, and aircraft), where substitution possibilities are much more limited. The non-Communist world's consumption of gasoline (the highest product and one used almost exclusively by automobiles and light aircraft) had been growing from 1950 to 1973 at 5.4 percent per year; for 1973 to 1984 that growth dropped to 0.4 percent per year. Consumption of middle distillates (diesel fuel and jet fuel for transportation, but also kerosene and home heating oil) grew by 7.7 percent per year in 1950 to 1973, but that growth dropped to 0.8 percent per year for 1973 to 1984. Use of heavy fuel oil (for industrial power and electric generation as well as some commercial heating) grew 7.0 percent per year in 1950 to 1973 but declined 3.6 percent per year for 1973 to 1984. Thus the entire drop in oil consumption from 1973 to 1984 was caused by the virtual loss of the fuel-oil market where substitution possibilities were greatest. The consumption of lighter products continued to grow, albeit much more slowly because of conservation, lower economic growth, and some substitution (diesel for gasoline, other fuels in home heating).

The greater demand elasticity for heavier products is important in understanding not only what did happen but also what is likely to happen under various price scenarios. On the one hand, it indicates what, at least in the short run, might set the bottom on oil price declines because at prices of $10 per barrel or lower, much of the fuel oil market could snap back fairly quickly (but be lost just as quickly if prices again rose). (This is good in that it limits the effect on non-OPEC producers of a price collapse while it retains the market check on sustainable price increases.) On the other hand, future oil consumption might be even less elastic at very high prices. Once almost all fuel oil use is stopped, substitution effects become much lower. (This is bad in that it tends to increase the effect of future disruptions or make them more difficult to offset with stockpile releases.)

Non-OPEC Oil. OPEC was squeezed not only by falling oil demand but also by the growth in non-OPEC oil supply. It is not, however, true that non-OPEC oil production spurted in response to the higher prices. Non-OPEC oil production grew from 1973 to 1986 at an average annual rate of only 3.3 percent per year, whereas it had grown in 1960 to 1973 at an average annual rate of 5.5 percent per year. Nevertheless, because non-OPEC oil production continued to grow while consumption essentially stagnated, OPEC oil production had to decrease by the amount of the increase in non-OPEC oil production. Some of the major new fields had already been found (e.g., Alaska North Slope and North Sea) and therefore cannot be completely attributed to the price increase. Their development was hastened, however, by the price rise and the perception of an energy crisis.

Oil production in non-OPEC less-developed countries could have increased much sooner, but many countries were reluctant to offer reasonable terms to attract oil companies to invest and drill; the oil price downturn in the 1980s made it apparent this was the only way they could develop their oil resources. Drilling costs rose rapidly along with oil prices, especially in the United States, as support industry capacities were reached and overextended. This, together with price controls and the excess profits tax in the United States, made oil supply less responsive to the price increases than it might have been. Drilling costs have since declined apace with prices. At today's drilling costs and oil prices there is every reason to believe non-OPEC production outside the United States will at least hold its own and probably increase moderately over the next several years.

Market Changes. The way oil is marketed also changed after OPEC wrested control of its oil production and pricing from the international "majors," the major integrated oil companies, which had dominated the market. The "third-party" market, where the majors sold crude in excess of the needs of their own refining and distribution systems to crude-poor companies, largely disappeared as their equity holdings were nationalized and OPEC members directly marketed their own crude. Long-term contracts came to account for a much smaller volume of the trade. Spot markets gained in volume and importance, not only for direct transactions but also as a basis for "market-related" contracts. New markets in oil futures and options were developed and grew spectacularly. These markets simultaneously provide a means for hedging against price risks and a way to speculate with less financial exposure. They also provide much greater price transparency than do traditional contract and spot markets because they are organized exchanges. During the 1986 oil price plunge, netback contracts played a major role in OPEC's attempt to regain market share. OPEC's December 1986 agreement to return to sales at official prices seems to have brought these to an end, at least for OPEC producers.

The oil market today is much different from that in 1973, and most of the changes seem to make the market less vulnerable to the type of crisis of the 1970s. Most countries and companies have greatly diversified their supply sources. There is considerably greater freedom to switch among suppliers. Most speculation goes on in the "paper barrel" futures and options markets where it does not as directly affect the physical "wet barrel" distribution of oil, only the price. And finally, the major consuming nations have built up substantial emergency stocks to use in any supply-threatening situation (stocks that did not exist in 1973), together with increased determination to use these stocks to prevent price shocks from getting out of hand.

Would OPEC Do It Again?

OPEC likely would not again even attempt to raise and maintain prices so far above free-market levels. In the 1973 and 1979 episodes there were real uncertainties as to the response to price increases of oil demand and supply. OPEC was not alone in misjudging them; most Western forecasters did also. But that uncertainty has now been removed. OPEC also has learned the disadvantages of a boom-and-bust approach. Too rapid gains are easy to waste, and contractions are difficult both economically and politically; steady growth produces more economic benefits and political stability.

There has always been a built-in tension in OPEC between those producers whose small reserves give them an interest in maximum short-run gains from high prices and those whose large reserves give them an overriding concern in preserving the long-run oil market. In 1973 and 1979, the price "hawks" carried too much weight in OPEC policy making. But the large reserve countries likely have learned this was not in their interest. The moderates, therefore, will try much harder to keep control in the future. As stated by a well-known energy analyst in a recent speech;

> In other words, it is quite plausible, though not inevitable, that during the 1990s the United States and the rest of the world's dependence on OPEC oil will once again rise to the level where OPEC can temporarily dictate oil prices at will. The organization succeeded twice in using temporary price explosions, caused by brief extraneous supply disruptions, to maintain oil prices for extended periods vastly above their free-market values. Thus the argument goes—and must be taken seriously—that under similar circumstances, or even without an extraneous trigger, OPEC would do so again. However, it must also be recognized that this proposition contains the facile assumptions that history repeats itself, that the 1990s will essentially resemble the 1970s, and that none of the major players on the supply or demand side has learned a lesson from the past. . . . Its [oil's] price will be determined less by market factors than by the cartel which controls its output. But if the cartel is to endure it must consider market factors in its pricing policy.[2]

Oil Supply Potential: Hubbert's Pimple Revisited

M. King Hubbert thirty years ago presented his pioneering use of logistic curves in oil discovery/production analysis (see Appendix 1 for details of Hubbert's hypothesis).[3] Application of Hubbert's logistic curve analysis to the world's oil production since 1918 on an annual basis results in the profile in Figure 4.4 (hence the nickname "Hubbert's Pimple"). This theoretical profile is based on an assumed total recoverable oil resource base of 2 trillion barrels, comprising

Figure 4.4. Hubbert's curve. World oil production profile (based on 2 trillion barrels. *Abbreviation:* mb/d, millions of barrels per day.

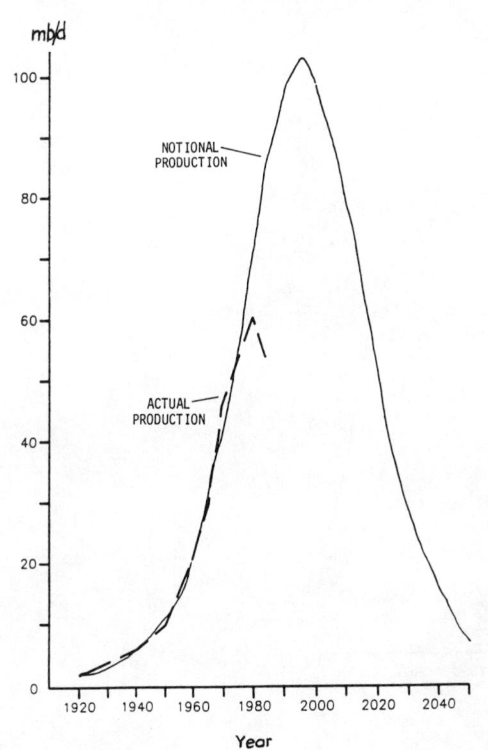

520 billion barrels already produced (end 1984), world-proven reserves on 690 billion barrels, and 790 billion barrels yet to be found.[4]

Hubbert's analysis provides a convenient way to put the OPEC episode in contest. His original analysis depicted the way oil production and consumption had been progressing before OPEC intervened and gives a background for looking at the consequences of OPEC. It also provides a means of keeping distinct the problems related to the existence of the cartel from those related to the eventual exhaustion of conventional oil resources.

Using a model based on Hubbert's analysis designed to forecast oil prices as a function of the difference between the steady state rate of growth of real income and the rate of change of oil discoveries, Leonidas P. Drollas presented a "plausible" scenario for future production as shown in Figure 4.5.[5] The shaded area between the original Hubbert curve and Drollas's notional curve represents the *potential*

Figure 4.5. Drollas's curve. World oil production profile (based on 2 trillion barrels).

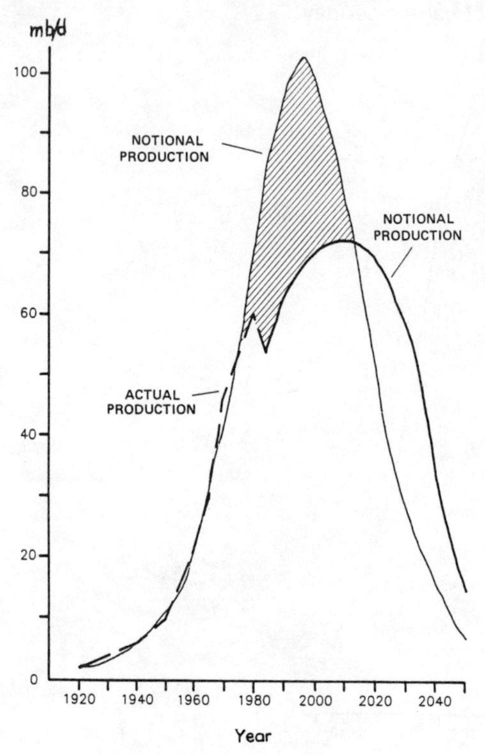

excess capacity (about 30 mb/d today) that can be expected to keep downward pressure on oil prices for three or four decades.

Inasmuch as almost half of the potential finds are estimated to lie in the Middle East, their rate of development is more of an OPEC policy variable than a purely economic or geological variable. But even if OPEC chooses to develop them more slowly to keep prices higher, the rate of development should rise for potential non-OPEC resources. Furthermore, the excess potential capacity within OPEC could make intracartel arguments over future quota levels intense.

Oil Price Stability

The argument made by some for the need for controls on the oil market rests as much on the desire for oil price stability as on the desire for any particular level of prices. Figure 4.6 and Table 4.4 show oil price history in terms of annual percentage changes. Over

ure 4.6. Year-to-year percentage change in the price of U.S. crude oil, 1901–1986.

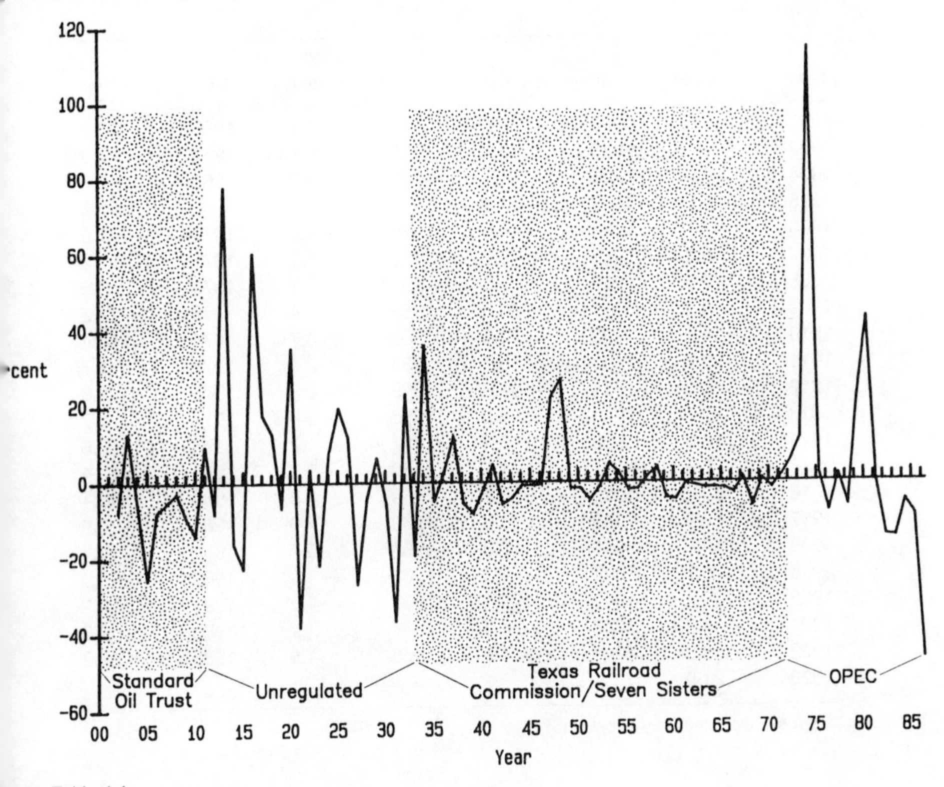

rce: Table 4.4.

the entire 1901 to 1986 period the average annual change in the real oil price was ±12.8 percent. Perhaps more interesting is a comparison among averages for the four different periods covered by the data.

1901 to 1911. During this period oil prices were reasonably stable under the Standard Oil Trust, with annual percent changes averaging ±9.4 percent. The average real price during this period was $12.46 (in 1985 dollars), and the range was from $8.32 to $17.57.

1912 to 1932. After the government broke up the Standard Oil Trust, oil markets were virtually unregulated. The average annual percent change during this period was much larger at ±22.9 percent, but the average real price was not too different at $13.46. The range was from $7.34 to $23.40.

Table 4.4. Year-to-Year Percentage Change in Real Oil Prices, 1902–1986

Year	% Change	Year	% Change	Year	% Change	Year	% Change
1902	−7.9	1924	8.0	1945	−1.1	1966	−3.2
1903	13.1	1925	19.3	1946	−0.8	1967	0.9
1904	−6.0	1926	11.7	1947	22.2	1968	−6.7
1905	−25.7	1927	−27.0	1948	26.7	1969	0.5
1906	−7.7	1928	−4.0	1949	−1.8	1970	−1.9
1907	−5.3	1929	6.2	1950	−1.9	1971	1.2
1908	−2.8	1930	−5.9	1951	−5.1	1972	5.5
1909	−9.3	1931	−36.8	1952	−1.4	1973	11.2
1910	−14.1	1932	22.9	1953	4.7	1974	114.4
1911	9.4	1933	−19.7	1954	2.1	1975	1.7
1912	−8.2	1934	35.8	1955	−2.3	1976	−8.0
1913	77.7	1935	−5.2	1956	−2.0	1977	1.7
1914	−16.5	1936	2.4	1957	1.2	1978	−6.6
1915	−22.8	1937	11.7	1958	3.7	1979	22.4
1916	60.3	1938	−5.7	1959	−4.5	1980	43.1
1917	17.5	1939	−8.6	1960	−4.9	1981	−0.9
1918	12.6	1940	−2.3	1961	−0.7	1982	−14.7
1919	−6.7	1941	4.4	1962	−1.3	1983	−15.0
1920	35.2	1942	−6.1	1963	−1.9	1984	−5.2
1921	−38.2	1943	−4.3	1964	−1.7	1985	−9.3
1922	3.3	1944	−1.0	1965	−1.9	1986	−47.2
1923	−21.9						

Source: Calculated from Table 4.1.

1933 to 1972. This period saw the most stability, first under the regulation of the Texas Railroad Commission and then under the Seven Sisters (the seven largest international majors—British Petroleum, Exxon, Gulf, Mobil, Shell, Texaco, and Standard Oil of California). The average annual percent change was only ±5.5 percent. The average real price was actually lower at $12.00, and the range only from $8.46 to $14.77.

1973 to 1986. In these, the OPEC years, prices again became highly volatile, with the average annual percentage change being ±21.7 percent. The average real price was much higher at $27.50, although prices fell to $14.27 by 1986 after reaching a peak of $43.84 in 1980.

Much of the price volatility in the earlier part of this price history arose because the market was relatively small. In the early years overall demand for oil was so small that discovery of a single supergiant oil field could flood the market.

The earlier periods, both controlled and uncontrolled, show a number of up-and-down cycles in response to changes in supply and demand—normal market behavior. The OPEC period, however, shows primarily two spikes related to the 1973–74 and 1979–80 crises, and then a sharp plunge when OPEC's attempt to hold prices at these inflated levels had killed demand and created a huge supply surplus. A freely working market would never have allowed such a large surplus to build up. Oil prices were much more stable in the Texas Railroad Commission/Seven Sisters and Standard Oil Trust periods when markets were controlled than in the uncontrolled period from 1912 to 1932. But prices have been highly unstable during the OPEC years despite (or perhaps because of) the cartel's efforts to control the market.

Both of the earlier control periods had average real prices not significantly different from, and actually lower than, the uncontrolled 1912 to 1932 period, which probably accounts in large part for the success of controls. One could say that the controllers largely accepted the market price level and concentrated on trying to reduce the frequency and size of deviations from this level. OPEC, on the other hand, tried to raise the price level (and in fact did double it on average), but thereby lost control and produced instability.

If the dominant OPEC members, especially Saudi Arabia, concentrate on trying to stabilize the market at a price closer to its historic level rather than push for maximum short-term gains, perhaps they will do better in the future. In any event OPEC is a fact of life in the absence of an international antitrust law and is likely to remain a major factor in oil pricing for decades. It holds the majority of oil reserves and potential oil resources. On balance, OPEC is more likely to do harm than good because of the cartel's desire to raise prices above the level a free market could produce.

How Much Price Stability Is Desirable?

Some degree of price volatility is normal and useful in any market. Prices serve as signals to both suppliers and consumers of changes they ought to make—to produce more or consume less of the product when its price increases, for example. An artificially stabilized price in a market that is moving out of balance runs the risk of allowing fundamental supply and demand to move seriously out of balance. On the other hand, a high degree of price volatility makes planning and investment decisions by both suppliers and consumers difficult. This is especially true in an industry such as oil, where producers' lead times tend to be long (five years or more from starting to search for oil until bringing it to market) and investments large (especially in frontier or offshore areas) and where major consumer decisions also involve capital investments rather than minor behavioral or taste changes.

The free-market price volatility of the 1912 to 1932 period does not, in retrospect, seem to have hurt. U.S. crude oil production grew 253 percent (6.5 percent per year) during the period, from 610,000 b/d to 2.151 mb/d. World production grew by 272 percent (6.8 percent per year), from 966,000 b/d to 3.590 mb/d. While data for the entire period are not readily available, the United States found 16.9 billion barrels of new proved reserves from 1918 through 1932, almost three times the proved reserves at the beginning of the period. Clearly the strong growth in demand and availability of economically developable resources outweighed whatever deterrent the volatile prices constituted.

Today the biggest threat to price stability is posed by the huge overhand of excess capacity, both developed and potential, created by OPEC's ill-advised attempt to maximize revenue by holding prices far above the level the market would allow. It is this threat, rather than the low price level (which is itself low only in respect to the exaggerated levels of the 1970s), that is the major deterrent to investment today. Prudent investors must consider the possibility that prices could again plunge to the level of mid-1986 or even lower if OPEC discipline cracks completely.

Conclusion

Rather than controls, reliance on the market (backed up by the insurance that strategic stocks and the International Energy Agency emergency program provide) may be the best approach to energy security in the face of deliberate or accidental disruptions of supply. Higher dependency on the Middle East increases the threat of accidental interruptions and temporary price shocks. Sustained higher prices, however, lead to lower dependency on the Middle East. It is not logical to fear getting higher dependency and sustained higher prices at the same time. Further, every barrel found anywhere in the free world outside the Persian Gulf reduces overall dependency on that region.

In all the concentration on import dependencies, one should not forget that most Middle East oil-producing countries rely on oil exports for some 90 percent of their foreign exchange and government budgets. They need to export oil even as the West needs to import it.

The United States already has been thoroughly explored for oil, and a large percentage of U.S. oil has been extracted. According to the U.S. Geological Survey, U.S. drillers now find only about 1/100 as much oil per discovery well as do drillers in the rest of the world.[6] Therefore, the survey deems it highly unlikely that the United States can satisfy its needs by stepping up exploration at home. An oil import fee could generate problems in the long run by exhausting limited domestic reserves faster, increasing demand in the rest of the world, and decreasing supply elsewhere.

OPEC, with its dominant position in world-proved oil reserves, is likely to be able to hold prices somewhat above the long-term historical level. But to do this, OPEC members must stem competition among themselves and keep production well below potential.

Appendix: Hubbert's Hypothesis

Hubbert's basic hypothesis was that cumulative oil discoveries follow a logistic curve over time. At first, when the oil market is in its infancy, discoveries are sporadic and small in size. Then follows an expansionary phase during which discoveries flow thick and fast and the market grows apace, with discoveries both leading and following the expansion of demand. However, as more oil is discovered over time, less oil is left, so oil discoveries level off at some stage and then decline inexorably. Hubbert further hypothesized that cumulative production follows a logistic cure that lags behind the discovery curve.

Although this is a geologically based model, not an economic one, the pattern it implies fits nicely with the behavior expected of a perfectly competitive oil market (so that the low-cost resources are produced first and producers do not hold oil off the market), costs that rise as oil discoveries become more difficult, and demand that is responsive to price increases and decreases.

The logistic curve in Figure 4.4 tracked world oil production reasonably well over the decades until the 1970s. When the twin crises of the 1970s caused oil prices to skyrocket and OPEC cut production to attempt to hold them at their inflated level, production fell badly below the theoretical curve. Normal market behavior was reversed, with some of the cheapest-to-produce oil shut in and the most expensive being produced as fast as possible. Furthermore, this curve implies that OPEC did not just shut in existing capacity but also failed to add capacity in line with what resources would have permitted and pure competitors would have done.

Notes

1. U.S. Department of Energy, *Energy Security: A Report to the President of the United States* (March 1987), p. 36.

2. John H. Lichtblau, "Future Course for U.S. Oil Supply, Demand and

Imports," speech delivered to Alternate Energy '87, 9 April 1987, pp. 2–3, 11.

3. M. King Hubbert, "Drilling and Production Practice," in *Nuclear Energy and the Fossil Fuels* (American Petroleum Institute, 1956), pp. 7–25.

4. Two trillion barrels is, for example, the median estimate given in Rand Corporation, *Giant Oil Fields and World Oil Resources* (June 1978), pp. 87–88.

5. Leonidas P. Drollas, "Oil Supplies and the World Oil Market in the Long Run," paper presented at the Eighth International Association of Energy Economists Conference, Tokyo, 7 June 1986, p. 15.

6. Charles D. Masters and David H. Root, "The Oil Outlook: A Realistic View," *Geopolitics of Energy* (March 1987): 2.

PHILIP K. VERLEGER, JR.

5. Structural Change in the 1980s: Effects and Permanence

THE structure of the world petroleum market was altered during the 1980s. The integrated framework was replaced by a more atomistic arrangement similar to that found in many other industries. This "restructuring" contributed to the growth of commodity market institutions. Simultaneously, the concentration of the production (i.e., share of the market controlled by a few producers) of crude oil dropped, creating the conditions that precipitated the price collapse.

The two structural changes had identifiable, independent impacts on the market, impacts which most analysts carelessly lump together. The breakdown of integration contributed to the development of commodity market institutions and ultimately led to the increase of price volatility. This change is probably permanent. The fall in concentration of crude production along with an associated increase in price sensitivity of consumers caused the decline in the level of oil prices. The decline in concentration may be transitory. Thus, while prediction is always risky, I conclude that commodity market institutions and price volatility are permanent features of the market while the regime of low prices may be transitory.

This chapter describes the principal changes in the oil market that occurred in the 1980s. The analysis begins by examining the altered role of the multinational companies. These firms controlled the flow of oil moving in international trade at the time of the 1973 embargo. However, their role was drastically reduced in the subsequent two decades.

The second part of this chapter examines the change in concentration in the world oil market. I argue that it is concentration, not surplus capacity, that is the relevant determinant of the level of oil prices and show that concentration declined during the 1980s. I suggest that prices may again increase in the future as concentration increases.

The final section of the paper describes the growth of oil commodity markets. These markets began to develop just as it was becoming apparent that OPEC could not or would not cut production as consumption declined. The price instability caused by OPEC's refusal to manage the market created financial risks that many oil companies could not, or would not, accept. As a result, the oil industry adopted instruments used in other commodity markets, including futures contracts, forward contracts, and market-related pricing. Firms not traditionally associated with the oil industry became active participants in the market. In one case a new entrant became a dominant player within four years.

The Diminished Role of the Majors

Change in the structure of the oil market began in the early 1970s, when many exporting nations seized producing interests previously held by the multinational oil companies. The nationalizations reduced the flexibility previously enjoyed by these firms while transferring responsibility for managing the market (maintaining a stable price environment) to the producing nations.

Nationalization reduced the size of the multinationals and cut their vertical channels of distribution. Simultaneously, the geographical diversification of the companies declined. In 1973 seven vertically integrated companies were involved in the production of crude oil in all OPEC countries and the sole or part owners of refining and marketing operations throughout the world. The scale of operations of these firms shrank by 1987, as may be observed from Table 5.1, which summarizes the volume of crude available to these companies, the refinery capacity owned by the firms, the volume of crude processed in their refineries, and their sales of products in 1973 through 1987.

The data in Table 5.1 show that the seven companies accounted for 29.3 mb/d (61 percent) of free world crude production in 1973 and processed 25.5 mb/d. Much of this crude oil was available from reserves actually owned by the companies while the remainder came from fields where the companies held long-term preferential arrangements (labeled preference crude in Table 5.1). The statistics presented in Table 5.1 also show that 80 percent of the multinationals' 1973 crude supplies came from the thirteen members of OPEC.

The position of the large multinationals changed dramatically between 1973 and 1987. The supply of crude available to them on a preferential basis declined steadily through the period. By 1987 the volume of crude had dropped by 64 percent from 29.3 mb/d to 10.5 mb/d. Further, the entire decline came from OPEC. Supplies from their former concessions in these countries fell from 23.5 mb/d to only 3.4 mb/d, an 85 percent reduction. These companies compensated for the reduced availability of crude oil by closing or cutting one-third of their refining capacity, reducing the volume of crude le 5.1. Principal Measures of Activity of the Traditional Major Oil Companies (in millions of els per day)

e available	*1973*	*1978*	*1981*	*1982*	*1983*	*1984*	*1985*	*1986*	*1987*
C preference	23.5	15.2	8.4	5.4	3.7	2.9	2.0	1.4	2.6
r	0	1.5	1.1	0.8	0.5	0.5	0.3	1.8	0.8
OPEC preference	5.8	6.4	6.1	6.2	6.8	7.4	7.3	6.8	7.1
Total	29.3	23.1	15.6	12.4	11.0	10.8	9.6	10.0	10.5
ing capacity	25.5	25.1	22.9	21.4	19.6	19.8	18.1	18.0	17.0
ery runs	23.4	20.2	15.6	14.1	13.3	13.7	13.0	13.4	13.2
uct sales	24.3	21.8	18.5	17.6	17.5	18.0	17.4	17.7	18.1
ourchases or sales	2.9	1.3	(2.9)	(5.2)	(6.5)	(7.2)	(7.8)	(7.7)	(11.0)

e: Petroleum Economics Limited.
The traditional major oil companies are British Petroleum, Chevron, Exxon, Gulf, Mobil, Shell, and Texaco.

processed, and trimming sales. Refinery runs were cut by 45 percent while product sales were reduced by 25 percent.

The multinationals shifted from net sellers of both crude and products to net buyers between 1973 and 1987. In 1973 crude oil supplies available to the seven multinationals exceeded their processing requirements by 3.8 mb/d. By contrast, in 1987 the volume of crude processed by the six remaining companies exceeded the volume of crude available to them by 2.8 mb/d. During the same period the companies' product deficit increased from 0.9 mb/d to 5.1 mb/d.

The change in scale of operations between 1973 and 1987 altered the relationship between the multinationals and other companies. Traditionally, the multinationals redistributed supplies to other companies through third-party sales. This role was abandoned during the 1980s as the volume of crude available to them declined from 62 percent of world supplies to 14 percent. Further, they became net *buyers* of a further 22 percent of world supply.

The diminished worldwide scale of the multinational companies is captured in Table 5.2. This table provides a comparison of crude availability, refinery runs, refinery capacity, and product sales of these companies to free world consumption. The changes in the market shares of these companies are particularly noteworthy since the level of consumption in 1973, 47.4 mb/d, was essentially the same as the level of consumption in 1987, 47.5 mb/d.

The downsizing of the multinational companies has been accompanied by a consolidation of their operations. The reduction in the sphere of operation of the companies in crude production is shown in Tables 5.3 and 5.4. These two tables display data on equity pro-

Table 5.2. Comparison of Crude Availability, Refinery Runs, Product Sales, and Refinery Capacity to Free World Aggregates

	Free World Demand (mb/d)	*% of Free World Demand of Traditional Major Oil Companies*			*Free World Refining Capacity (mb/d)*	*Majors Refining Capacity as Share of Free World Capacity (%)*
		Crude Available	*Refinery Runs*	*Product Sales*		
1973	47.4	62	49	51	50.3	51
1978	50.3	46	40	43	63.5	40
1981	46.5	34	34	40	65.1	35
1982	45.1	27	31	39	62.3	34
1983	44.9	24	30	39	59.3	33
1984	45.5	24	30	40	57.6	34
1985	45.3	21	29	38	56.4	32
1986	46.7	21	29	38	56.1	32
1987	47.5	22	28	38	56.2	30

Source: BP Statistical Yearbook, Table 3.1.

duction of the seven multinational companies throughout the world in 1969 (Table 5.3) and 1985 (Table 5.4). The first of these tables, Table 5.3, is reproduced from M. A. Adelman's seminal study *The World Petroleum Market.*[1] Adelman originally introduced Table 5.3 and several similar tables for earlier years to illustrate the nature of joint relationships between the multinational companies in OPEC. The diversification and joint nature of these operations permitted the companies to effectively redistribute supplies during the Arab embargo.

The nature of joint operations is clearly noted in the data for 1969. The companies were able to rely on a wide range of supply alternatives during the years before and immediately after the Arab embargo. The seven major companies also held a favored position vis-à-vis other oil companies at this time because they controlled access to 94 percent of OPEC oil in these years.

By 1985 the multinational companies had lost their access to equity production from OPEC nations. Indeed, comparison of the upper half of the tables between 1969 and 1985 is quite remarkable. During this period OPEC production declined from 20 mb/d to 17 mb/d. During the same period the equity production by the multinational companies in OPEC nations declined from 80 percent of total OPEC output (15.8 mb/d) to 4 percent of total OPEC output (0.6 mb/d).

The reduced diversity in supply may be noted by comparing Tables 5.3 and 5.4. Each company lost all or most of its supplies from OPEC nations. The process began with nationalization by Libya before 1973 and was followed by Kuwait's seizure of the assets owned by British

ole 5.3. Equity Production of Principal Crude Oil Companies and Producing Countries, 1969
usands of barrels per day)

	Exxon	*Mobil*	*Texaco*	*British Petroleum*	*Shell*	*Chevron*	*Gulf*	*CFP*	*Others**	*Total Production*
EC										
di Arabia†	870	289	870	—	—	870	—	—	—	2,899
ait	—	—	—	1,254	—	—	1,254	—	—	2,508
eutral Zone	—	—	—	—	—	—	—	—	—	0
ed Arab Emirates										
ou Dhabi	49	49	—	236	99	—	—	167	21	621
ubai	—	—	—	—	—	—	—	—	—	—
	208	208	208	1,192	417	208	208	178	394	3,221
	181	181	—	361	361	—	—	361	76	1,521
ar	23	23	—	45	203	—	—	45	9	348
ezuela	1,472	112	171	—	916	56	383	—	424	3,534
on	—	—	—	—	—	—	—	—	—	—
ria	—	—	—	168	168	—	200	—	—	536
nesia‡	25	25	285	—	138	285	—	—	90	848
ria	—	—	—	—	85	—	—	249	602	936
ador	—	—	—	—	—	—	—	—	—	—
a	683	126	181	155	119	181	—	—	1,561	3,006
Total OPEC	3,511	1,013	1,715	3,411	2,506	1,600	2,045	1,000	3,177	19,978
I-OPEC										
lle East	—	—	—	—	256	—	—	—	498	754
h America	—	—	—	—	—	—	—	—	—	—
h America	—	—	—	—	—	—	—	—	—	—
pe	—	—	—	—	—	—	—	—	—	—
a	—	—	—	—	63	—	36	—	406	505
ralasia-Asia	—	—	—	—	—	—	—	—	—	—
Total§	3,511	1,013	1,715	3,411	2,825	1,600	2,081	1,000	4,081	21,237

ce: M. A. Adelman, *The World Petroleum Market* (Baltimore: Johns Hopkins University Press, 1972), pp. 80–81.
* Smaller oil companies and national oil companies. CFP, Compagnie Française des Pétroles.
di Arabia includes Bahrein.
onesia includes Brunei and Sarawak.
ere are numerous small discrepancies in the data source, and it is impossible to reconcile them fully; hence the failure
ail to add to totals is not wholly due to rounding." Including small self-sufficient countries and production not identified
untry, the total non-Communist production outside of North America was 24,400.

Petroleum (BP) and Gulf. Saudi Arabia, Abu Dhabi, Nigeria, and Venezuela quickly copied these two countries, imposing total or partial nationalizations. Concessions in Iran were terminated in 1979 as part of the general upheaval there. As a result, the six remaining multinational companies were left with essentially no stake in OPEC nations by 1985.

The multinational companies compensated for the decline in crude available to them and the reduced diversity in their supplies by shrinking their sphere of operations. They closed or sold approximately

Table 5.4. Production of Principal Crude Oil Companies and Producing Countries, 1985 (in thousands of barrels per day)

	*Exxon**	*Mobil*	*Texaco*	*British Petro-leum*	*Shell*	*Chevron†*	*Others‡*	*Total Produc-tion*
OPEC								
Saudi Arabia	—	—	—	—	—	—	3,565	3,565
Kuwait	—	—	—	—	—	—	920	920
Neutral Zone	—	—	54	—	—	—	286	340
United Arab Emirates								
Abu Dhabi	19	—	—	68	33	—	770	890
Dubai	—	—	—	—	—	—	390	390
Iran	—	—	—	—	—	—	2,215	2,215
Iraq	—	—	—	—	—	—	1,440	1,440
Qatar	—	—	—	—	—	—	340	340
Venezuela	—	—	—	—	—	—	1,730	1,730
Gabon	—	—	—	—	18	—	137	155
Nigeria	—	—	9	—	150	89	1,228	1,475
Indonesia	7	42	23	—	—	20	1,243	1,335
Algeria	—	—	1	—	—	—	969	970
Ecuador	—	—	72	—	—	—	203	275
Libya	—	—	—	—	—	—	1,105	1,105
Total OPEC	26	42	158	68	201	109	16,541	17,145
NON-OPEC								
Middle East	—	—	—	—	133	—	667	80[illegible]
North America	884	352	742	783	578	101	8,919	12,36[illegible]
South America	18	83	103	—	—	7	4,470	4,68[illegible]
Europe	434	138	170	510	456	115	2,138	3,96[illegible]
Africa	—	—	16	—	225	79	1,270	1,59[illegible]
Asia-Australasia	330	134	5	—	200	6	1,280	1,95[illegible]
Total	1,692	749	1,195§	1,361§	1,793	416§	35,284	42,49[illegible]

Sources: BP Statistical Review of World Energy (June 1987); *Financial and Operating Information 1983–19*[illegible] British Petroleum 1988; Annual reports and correspondence with company officials.

* Includes gas plant liquids; excludes Canadian oil sands production.

† Includes Gulf Oil.

‡ Smaller oil companies and national oil companies.

§ Includes production not identified by country.

one-third of their refineries between 1973 and 1987, effectively conceding market share to other firms, including the oil companies established by oil-exporting nations. The overall decline in the ownership of refining capacity is shown in Table 5.2, where data on the share of worldwide refining capacity and non–U.S.-refining capacity owned by the multinational companies are displayed. On a worldwide basis, the share of capacity owned by the multinationals declined from 47 percent in 1973 to 30 percent in 1987.

le 5.5. Refining Capacity, Largest Companies' Proportion by Area, 1966 (in thousands of rels per day)

	Total	*8 Largest Companies*	%	*Exxon*	*Mobil*	*Shell*	*SoCal*	*Texaco*	*Gulf*	*British Petroleum*	*CFP*
				%							
sian Gulf	1,662	1,363	82	7	4	4	13	13	10	30	2
er Middle East	293	117	40	1	16	8	4	4	4	2	2
h Africa	303	5	2	5	1	3	0	0	0	1	4
er Africa	402	274	68	3	12	20	5	6	0	20	2
an	2,211	480	22	5	3	5	4	4	0	0	0
er Asia	986	345	35	14	5	7	3	3	2	1	0
tralia, New Zealand, Malaysia	758	632	83	6	14	26	7	7	0	22	0
pean Community	9,527	5,888	62	18	4	17	2	3	1	12	5
America	4,283	2,718	63	27	2	21	1	9	3	—	—
Total world	20,425	11,822	58	16	4	15	3	5	2	9	3

ce: M. A. Adelman, *The World Petroleum Market* (Baltimore: Johns Hopkins University Press, 1972), p. 96.

le 5.6. Refining Capacity, Largest Companies' Proportion by Area, 1975 (in thousands of els per day)

	Total	*8 Largest Companies*	%	*Exxon*	*Mobil*	*Shell*	*Texaco*	*British Petroleum*	*CFP*	*SoCal*
				%						
ian Gulf	2,591	1,451	56	9	7	7	6	14	1	11
r Middle East	683	148	22	8	6	2	2	2	1	1
h Africa	481	16	3	0	1	0	0	0	1	—
r Africa	906	586	65	3	15	22	5	11	6	4
n	5,417	1,809	33	3	6	8	8	—	—	8
r Asia	3,245	1,254	39	7	5	20	2	1	—	2
ralia, New Zealand, Malaysia	884	614	69	9	11	18	6	20	1	4
pean Community	20,138	10,450	52	13	5	12	4	22	6	2
America	7,005	2,967	42	20	1	13	7	—	0	1
Total world	41,349	19,295	47	12	5	12	5	7	3	3

ce: *International Petroleum Encyclopedia* (Petroleum Publishing Co., 1976).

The reductions in refining capacity and the associated withdrawals from marketing had the effect of reducing the multinational companies' geographical diversification, just as the nationalization of their producing assets had reduced the companies' diversification of crude sources. This pattern can be observed from Tables 5.5 to 5.7, which show the market shares of the largest companies by region of the world for 1966, 1975, and 1987.

Table 5.5, also reproduced from *The World Petroleum Market*,

Table 5.7. Refining Capacity, Largest Companies' Proportion by Area, 1987 (in thousands of barrels per day)

				%					
	Total	*8 Largest Companies*	*%*	*Exxon*	*Mobil*	*Shell*	*Chevron*	*Texaco*	*British Petroleum*
Persian Gulf	3,084	310	10	0	3	4	2	2	0
Other Middle East	1,208	98	8	0	4	2	0	0	2
North Africa	1,391	4	0	0	0	0	0	0	0
Other Africa	1,975	477	24	1	4	6	3	4	5
Japan	4,790	306	6	3	0	4	0	0	0
Other Asia	4,619	1,246	27	5	4	9	3	3	2
Australia, New Zealand, Malaysia	892	641	72	3	11	32	4	4	16
European Community	13,935	5,630	40	11	1	10	1	4	8
Latin America	6,873	436	6	2	0	2	1	2	0
Total world	38,767	9,149	24	6	2	7	1	3	4

Source: International Petroleum Encyclopedia (Petroleum Publishing Co., 1988).

shows that the eight largest companies owned 59 percent of non–North American-refining capacity and held 22 percent of Japanese capacity; 83 percent of capacity in Australia, New Zealand, and Malaysia; and 62 percent in Europe. Further, Adelman notes that the eight multinationals had the right to supply up to 51 percent of Japan's crude requirements through their joint ownership of Japanese refineries with local companies.[2]

The effect of the decline in ownership after 1966 may be noted by comparing Table 5.7 with Table 5.5. The multinationals owned only 6 percent of Japanese refining capacity in 1987 and 40 percent of refining capacity in western Europe. Further, many of the companies that had formally been involved in these regions had ceased all operations. Shell, Chevron, and Texaco had withdrawn from Japan while Mobil and Texaco had reduced their involvement in Europe. During this period every one of the majors had withdrawn from one or more redundant European countries.

Competition among buyers for crude oil increased as the multinational companies withdrew from one region after another. Most of the refineries were not closed but sold, often to the firms that had previously operated the units jointly with the multinational company selling the unit. The termination of the joint refining operation, particularly in Japan, thrust the operator into a new role as a buyer of crude. Initially supplies were obtained through third-party contracts. Under these arrangements multinational companies with access to OPEC crude oil resold crude to buyers such as Japanese refiners under long-term contracts at prices that were linked to OPEC official selling prices. Most third-party contracts were eliminated, however, during the Iranian crisis. BP canceled its sales in January after it lost its access to Iranian crude. Exxon followed BP's lead, announcing that it would phase out third-party sales over the next year.[3]

The actions of Exxon and BP forced third-party buyers into the market at a time when supplies were tight, effectively reducing buyer concentration and providing the ideal environment for oil-exporting nations to raise prices sharply. The oil-exporting countries seized the opportunity.

The changes described on the previous pages can be summarized in terms of concentration. The nationalizations of the producing interests of the multinational companies caused concentration among buyers of crude oil to decline. Under a classic competitive model, the reduction in buyer concentration should enable producers to raise prices, and crude oil prices did, in fact, increase. However, the initial increases pushed prices well past the competitive level as the members of OPEC attempted to find the price a monopolist might charge.

Mismanaging the Oil Market: OPEC's Failure

The responsibility for maintaining a stable price environment fell to oil-exporting nations when they nationalized the reserves of the multinationals. Prior to the nationalizations price stability had been assured by the multinationals. Adelman credits this stability to the steady growth in demand, limits on imports imposed by consuming nations, and the willingness of the multinational companies to restrain production.[4]

The increase in prices after 1973 eliminated the economic incentive that had previously motivated companies to maintain a stable price environment. Exporting nations were forced to accept the responsibility for stabilizing prices. Initially they responded to the pressure of fluctuating demand by adjusting production.

The willingness to vary output enabled OPEC to sustain the threefold increase in prices. OPEC's operations during the 1970s were precisely those that are prescribed for a cartel. The organization established a price and adjusted output to defend it. However, the high prices had the predicted impacts on both demand and the "competitive fringe."[5] Conservation and substitution were encouraged. Between 1979 and 1984 free world consumption declined by 10 percent. High prices established in 1973 and defended through the 1970s also stimulated increased production from the competitive fringe. Discoveries in Alaska and the North Sea were developed. New discoveries were made in Mexico and quickly brought into production.

Exploration efforts in Malaysia, Egypt, Angola, and other areas were accelerated.

The exploration effort was also spurred on by a desire of the multinational oil companies to replace the reserves they had lost through nationalizations. This effect may be noted by comparison of Table 5.3 with Table 5.4. Exxon increased its European production from nothing to 434,000 barrels/day between 1969 and 1985 and raised output in non-OPEC Asian countries from 0 to 330,000 barrels/day. Shell recorded an increase in production in Europe from 0 to 456,000 barrels/day while raising its production from non-OPEC African countries to 200,000 barrels/day from 0. BP, Mobil, Texaco, and Chevron all recorded similar increases.

The increased production on the competitive fringe forced OPEC to institute further production cuts during the mid-1980s. Output declined from 32 mb/d in 1980 to 17 mb/d by 1984. The members, faced with the decline in demand, made repeated tries to organize output reductions, reductions which generally failed. Ultimately, the members of OPEC were forced to abandon their use of official prices because they could not cope with the lower level of demand. The members also temporarily suspended quotas on production, staging a disastrous war for market share in 1985 and 1986.

The price collapse is explained by the decline in the concentration of crude production. Through the early 1980s the dominance of Middle Eastern oil-exporting nations fell as consumption dropped and production from non-OPEC sources increased. This change is normally described by reference to surplus production capacity in OPEC (see, for example, Chapter 3). However, economists (and, more important, the regulators of competition such as the Federal Trade Commission and the Department of Justice) recognize that it is market share, not capacity, that is relevant to the ability of a cartel or firm to maintain prices above competitive levels.

The level of concentration is measured here using the Herfindahl-Hershman index, or HHI. The HHI is computed as the sum of the squares of the market shares of each of the competitors and is thus distributed between 0 and 10,000. Higher values of the index are associated with more concentrated markets. For example, the index is 10,000 for a monopoly market. The U.S. Department of Justice (DOJ) has formally stated that the HHI is to be used in determining whether a merger of two firms will be permitted or opposed. In general, DOJ will take no action if the premerger HHI is less than 1,200 and the postmerger will increase by less than 50 to 75 points. Mergers will receive greater scrutiny as the HHI increases, and almost all mergers will be opposed if the premerger HHI exceeds 1,800.

An HHI index for market shares of oil traded in international markets appears to confirm the DOJ findings. An imperfect relationship between the concentration in the international oil market and the level of crude oil prices has existed since 1970. A representation of the relationship between prices and concentration is shown in Figure 5.1. Examination of Figure 5.1 shows that the 1973 and 1979–80 price increases are associated with sharp increases in concentration while the 1986 decline is related to a sharp drop in concentration. The index of concentration shown in Figure 5.1 is calculated from composite data on the market shares of the multinational oil companies and oil-exporting nations. Company data were used for the period from 1969 to 1973 while data for oil-exporting nations were used from 1974 onward.

The approach to the calculation of market shares assumes that the responsibility for determination of output shifted from the companies to exporting nations in 1974. Prior to 1973 the multinational companies managed production facilities in exporting nations and determined output levels. After 1973, decision-making responsibility was shifted to the oil-exporting nations.

It may be observed that the 1974 shift caused a sharp increase in the concentration of the international oil market. The HHI for in-

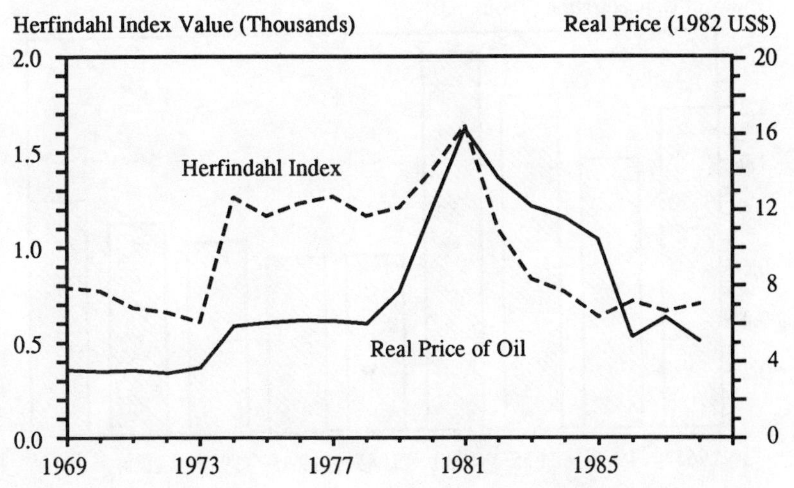

Figure 5.1. International oil market index of concentration compared with real price of oil.

Sources: Charles River Associates, 1989; *Petroleum Economics Monthly*, November 1989; Philip K. Verleger, Jr.

ternationally traded oil increased from 604 to 1,265 in one year. This increase coincided with the threefold increase in oil prices. The second sharp increase in concentration occurred in 1979 during the Iranian crisis. This increase is the direct result of the rise in Saudi Arabia's market share from 16.9 to 22.4 percent of the world market. Once again, the increase in concentration was accompanied by a tripling of oil prices. The price collapse of the 1980s was also associated with a change in concentration. By 1986 the HHI and real prices had dropped back to levels that had prevailed prior to 1974. This analysis shows that the 1986 oil price collapse resulted from a decline in market concentration. Increased production from non-OPEC sources combined with a decline in use to increase competitive conditions in the world market.

This trend toward a more competitive crude oil market may be reversed in the next decade. Market concentration in the 1990s will depend upon a number of factors. Among the most important are the level of demand, the level of non-OPEC production, and, obviously, the distribution of that output. For illustrative purposes, indices of concentration were calculated using the low-price case estimates of supply and demand published by the U.S. Department of Energy (DOE) in its 1989 *International Energy Outlook*. The projections shown in Figure 5.2 suggest that concentration in the export

Figure 5.2. Index of world oil-exporter concentration (low-price Department of Energy case with low Mideast exports).

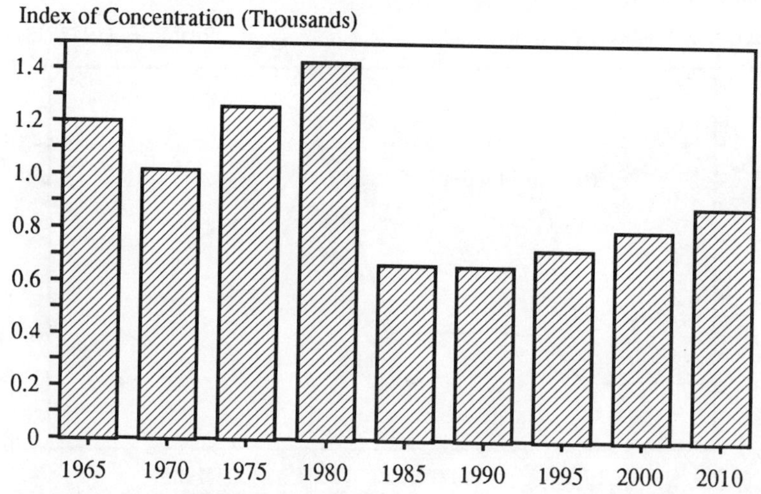

Sources: Charles River Associates, 1989; *Petroleum Economics Monthly*, November 1989; Philip K. Verleger, Jr.

market will not increase dramatically over the next twenty years. A logical conclusion of this result is that oil prices will tend to remain at present levels rather than rising dramatically.

However, such a conclusion may be premature. Production from non-OPEC sources such as the United States or the Soviet Union could decline at higher than anticipated rates or consumption could increase at a rate above the level projected by DOE. Either outcome would lead to a higher level of concentration and higher prices. Thus, the conclusion from this exercise is simply that concentration of markets determines price levels, and the decline in prices in the 1980s is linked to the drop in concentration that occurred after 1981.

The Growth of Oil Commodity Markets

Oil commodity markets developed to address a second problem created by the breakdown of integration: the exposure of buyers of crude oil to significant risks created by price fluctuations. This section explains the developments behind the growth of these new markets and then presents estimates of their size.

Oil and Commodity Markets. The economic literature does not provide a definition of a commodity. However, petroleum markets did not share those characteristics normally associated with commodities prior to 1980 but did by 1988. The most obvious manifestations of change are to be found in the development of large, liquid futures and forward markets. Less obvious signs of change are to be seen in the character of the participants in the market, the increase in price volatility, the role of stocks, and the changing nature of price formation.

Hieronymus lists three characteristics common to successfully traded commodities:

> First, they are all bulk commodities that can be described and the separate lots of which are more or less interchangeable. Second, none of the commodities has been processed or manufactured to the point of being a product identified with the process of a particular firm; in general they are bulk commodities but are not, in a strict sense, raw materials. Third, prices are variable and relatively competitively determined.[6]

Baer and Saxon offer an even more precise definition of commodities. First, the commodity must be sold in homogeneous units. Second, it must be susceptible to standardization. Third, supply and demand must be large so that speculators with large financial resources cannot gain control over supply or demand. Fourth, the supply of the commodity should flow naturally to the market without being restricted by artificial restraints imposed by either governments or private organizations. Fifth, supply and demand must be uncertain. Finally, the commodity must be storable so that forward and futures contracting can take place.[7]

The structural changes in the oil market described in the previous two sections of this chapter contributed to the development of oil commodity markers by enabling oil to meet these criteria. The process began with the nationalizations of the interests of the multinational companies by OPEC nations. The evolution of oil into a commodity was then accelerated by the shrinkage of the multinational companies. Finally, the transformation was completed when OPEC failed to maintain a stable price environment. Three other factors also contributed to the growth of oil as a commodity: the adoption of specific resource tax regimes in consuming nations, the decline in consumption, and the removal of government regulations.

The changes described earlier in this chapter affect the fourth of Baer and Saxon's six criteria, the flow of oil to the market. Nationalizations of the interests of the eight majors doubled or tripled the number of sellers of crude and removed the control multinationals had exercised over the volume of production. The withdrawal of the multinationals from third-party sales agreements then forced oil-exporting nations to market an increased share of their crude while creating a ready market among the former third-party buyers. Finally, the development of new resources outside of OPEC in countries such as Norway, the United Kingdom, Mexico, Egypt, and Colombia added to the diversity of supply, reducing any impediments to the flow of oil to the market.

The development of oil commodity markets was also boosted by the removal of regulations that had previously encumbered the industry. The principal change occurred in the United States. Controls had been imposed on the U.S. oil industry for more than twenty years. The nature of the controls became progressively more restrictive over the period, and from 1973 to 1981 the U.S. industry operated under a system that required mandatory distribution of crude and product at prices set by regulation, not the free market. Governments of two consuming nations, the United States and the United Kingdom, gave a further boost to the development of commodity markets by imposing new taxes on crude oil, taxes that created a higher marginal rate for production from resources than other activities such as refining and marketing. The more pernicious of these was the Petroleum Revenue Tax (PRT) imposed in 1978 by the United Kingdom.

The effect of the PRT was to raise the effective, as opposed to the published, tax rate on oil production. At the margin the rate may have actually exceeded 100 percent. Producers responded to these circumstances by diverting their crude to the spot market while simultaneously acquiring supplies for their refineries from other sources. These purchases and sales permitted them to establish their own arm's length price (the price they received for the oil), thereby avoiding the excessively high and unintended marginal tax rate. The sales by producers of North Sea oil also had the effect of increasing the flow of oil to the market.

Another factor that contributed to the growth in oil commodities markets has been the increase in uncertainty concerning the level of demand. (It may be recalled that Baer and Saxon list uncertainty as to demand as a prerequisite for the successful creation of a commodity market.) Prior to 1973 the growth in demand for oil had been predictable. The easy predictability of oil consumption ended, however, with the price increase in 1973. Consumption began to vary erratically as economic growth slowed and higher prices induced conservation and fuel substitution during the period from 1982 to 1986.

The increased uncertainty as to demand along with the deregulation of the oil market, adoption of resource taxes such as the PRT, and the reduced scope of operation of the multinational oil companies all have contributed to the growth of oil commodity markets. In fact, oil provides almost a textbook demonstration of the principles listed by Hieronymus. Commodity markets sprang to life when the market was deregulated, after demand became uncertain, and because oil began to flow freely to the market.

This transformation would seem to be permanent. The trend to reintegration will not stifle the free flow of oil to the market; in fact, it may increase the flow. Downstream integration will also not lead to a reimposition of regulation. The move to integrated operations will do nothing to eliminate the uncertainty as to demand. Thus, commodity markets are likely to remain an integral part of the oil industry for the foreseeable future.

Types of Commodity Markets. Three types of markets are traditionally associated with commodities: spot, forward, and futures. All three markets share certain common characteristics, but each has its own unique characteristics. The principal common characteristic is that a transaction on each market involves the actual or contemplated exchange of a commodity for cash *with no further consideration*. By contrast, a term sale, third-party sales, or long-term contracts between buyers and sellers involve the prospective exchange of specific volumes of the commodity for cash over a period of time under pricing terms that are specifically established in the contract. Obligations in these three markets are binding in that the trader must perform unless the contract specifying the action is transferred to another party.

Transactions on spot markets generally involve the immediate exchange of individual lots of the commodity for cash. The transactions are between principals, and actual title to an identifiable lot or quantity of the commodity is exchanged. Further, payment for delivery is either made immediately or within a relatively short period of time. Spot market transactions may involve the exchange of standardized units meeting uniform specifications (for example, barge lots of heating oil meeting specific specifications). However, many specific customized deals where either the volume and/or the specification of the deliverable commodity will differ from generally prevailing standards also occur on spot markets.

Transactions on forward markets, like transactions on spot markets, are between principals. However, the exchange is deferred, as is payment. Thus a specific buyer will agree to accept a volume or amount of the commodity from a seller at some date in the future *at a price that is determined at the time of the transaction*. Normally, payment for the commodity does not occur until delivery takes place. (In legal terms, transactions on forward markets involve the sale of a "cash commodity for deferred shipment or delivery.")[8] Over time forward market transactions may become standardized. Additional transactions in the same contract between principals will occur as the contract becomes standardized. Thus "daisy chains" may develop where A sells a forward contract to B who then sells it to C, etc. Clearing of contracts on forward markets through offset or by payment of differences will increase as standardization increases. This payment of differences representing the profit or loss on a transaction to a given trader will become common because it is more efficient and less costly than the actual physical exchange of the commodity.

Transactions on a futures market share similarities to those on well-developed forward markets. The commodity is bought or sold for future, as opposed to immediate, delivery. The lot is highly standardized and profits and losses are often settled in differences. However, there are also several important differences between futures and forward contracts.

First, futures contracts are bought and sold on organized, regulated exchanges through auction (usually via open outcry) whereas forward contracts are negotiated between principals. Second, transactions are anonymous—the buyer and the seller do not know one another. Indeed, the clearinghouse is technically the seller to every buyer and the buyer to every seller. This provision of anonymity is thought to make futures markets more competitive and more "fair." While traders in forward markets may discriminate among their various competitors, offering better terms to one than another based on various criteria such as perceived creditworthiness, past performance, etc., traders on the futures markets, on the other hand, theoretically receive the identical treatment.

Futures and forward markets also differ in the determination of the delivery price. The delivery price for a futures contract held to maturity is established when the contract expires, not at the time of purchase. Futures transactions generally involve the payment of initial and variational margins, whereas deposits are generally not required for forward transactions. Finally, transactions are cleared through a clearinghouse rather than between principals to the transaction.[9] Table 5.8 provides a summary of the individual characteristics of the three types of contracts.

Futures and forward contracts will also differ in terms of the delivery arrangements that apply to the contracts. Forward contracts may often offer flexible delivery arrangements that contemplate the specific details that will be worked out between principals. Futures contracts will, on the other hand, list very explicit and precise delivery terms.[10]

Oil Commodity Markets. Spot transactions in petroleum markets have been a regular part of the industry throughout its history. The most well-known and well-defined spot market has been the Rotterdam product market. Smaller product markets also existed on the U.S. Gulf Coast, on the East Coast, and in Singapore.

Forward markets in petroleum began to develop in the 1970s. The oldest and largest is in Brent, a crude produced in the U.K. sector of the North Sea. Additional forward markets have developed in Dubai (a Middle East oil similar to Arabian Light, the crude produced by Saudi Arabia), Alaskan North Slope crude, heating oil delivered to Boston, residual fuel oil delivered to the United Kingdom, Russian gasoil (heating oil) delivered into Europe, and gasoline and "open spec." naphtha delivered to Japan. Futures markets were created at the same time. The first successful futures markets in the petroleum industry were created in 1978 in New York and London.

The Size and Growth of Oil Markets. Petroleum commodity markets were small as late as 1980, according to any measure of size one chooses to use. By 1988, however, they had become very large. The unimportance of the market in 1980 was noted by Roeber, who commented, "The NWE [North West Europe] and Med [Mediterranean]

Table 5.8. Comparison of Significant Elements of Spot, Forward, and Futures Transactions

	Cash or Spot	*Forward*	*Futures*
Public trading in central marketplace	No	No	Yes
Specific trading hours	No	No	Yes
Settlement by delivery	Almost always	Rarely	Rarely
Price transparency			
Transactions publicly reported	No	No	Yes
Competitive or open outcry	No	No	Yes
Public disclosure	If by one or both parties	If by one or both parties	Yes
Guarantor of performance	Individual parties	Individual parties	Clearing parties
Initial margin deposits	No	Rarely	Always
Letters of credit	Often	Seldom	On delivery
Intermediaries (broker)	Infrequent	Infrequent	Usually
Daily margins (marked to market)	No	No	Yes
Public participation	Infrequent	Infrequent	Substantial
Contract terms			
Standardized	No	Optional	Yes
Fixed quantity per contract	No	Often	Always
Standard quality	No	Often	Always
Fixed delivery locations	No	Often	Always
Standard delivery period	No	Often	Always
Force majeure term	Sometimes	Sometimes	Standard
Regulated	No	Rarely	Yes

Sources: Jeffrey Williams, *The Economic Function of Futures Markets* (Cambridge: Cambridge University Press, 1986); Committee on Commodities Regulation of the Association of the Bar of the City of New York.

spot markets are small in relation to the total trade in oil products in Europe. They serve the function of balancing supply and demand at the margin."[11] Roeber estimated that the average volume of trade amounted to roughly 1 mb/d at that time. Estimates of the current size of the European product market are not available, nor are estimates of the size of the various U.S. spot markets. However, the absence of detailed reports and the very function of the markets—as a balancing mechanism—would imply that the total volume of trade has remained roughly constant. Total volume is probably less than 5 mb/d.

Calculation of the size of the spot market for crude oil is more complicated than product markets. Transactions are not concentrated in a few geographic locations, confined to a limited variety of oils, and of a uniform size. Title can change hands literally any place in the world and crude can be delivered anywhere. Deals in more than one hundred different types of crude oil are reported every year. Transactions range from very small pipeline batches of a few thousand barrels to cargos of more than 4 million barrels carried on ultralarge crude oil carriers.

A rough estimate of the size of the cargo market can be developed using data published by *Petroleum Argus*. These data indicate that the number of transactions in the spot market increased from 54 in 1973 to 3,058 in 1985. Depending upon the average size of the ship, volumes could have ranged from 27 to 150 million barrels in 1973 and from 1.5 billion barrels to 4.5 billion barrels (4 mb/d to 12 mb/d) in 1985. We would estimate that the volume of oil moving in the spot market was 50 million barrels in 1973 (150,000 barrels/day or 0.3 percent of free world consumption) and roughly 3 billion barrels (8 mb/d or 18 percent of free world consumption) in 1985. This estimate is based on an assumption that the average-size cargo was in the range of 150,000 to 180,000 dead weight tons or roughly 1 million barrels. When combined with transactions in products, it appears that the total volume of trade on the spot market was between 11 and 17 mb/d in 1988.

Trading on spot markets is supplemented by a much larger volume of trade on forward markets for crude petroleum and products. The Brent market is the largest of these. Brent is one of several streams of crude oil produced in the North Sea. A blend of several types of crude, output amounts to 900,000 barrels/day approximately 50 percent of total North Sea production (and supplying 4 percent of world consumption).[12] Production from fields producing the Brent blend is moved to Sullom Voe in the Shetland Islands by pipeline where it is delivered onto tankers. By custom Brent production has been sold in lots consisting of 500,000 barrels[13] or approximately fifty-four deliverable cargos per month.

Mabro reports that trading in Brent crude began in 1981 with short selling and covering by a number of traders, noting that "in many cases short selling also meant forward selling since traders were often trying to balance rather limited North Sea f.o.b. availabilities against c.i.f. arrivals of long-haul crudes from the Middle East."[14] Active trading by producers began in late 1981 or early 1982 as companies sought to limit tax liability under the PRT.

The number of reported transactions in the Brent forward market has increased from 700 in 1983 (950,000 barrels/day) to 8,000 in 1988 (13,150,000 barrels/day), making the Brent market almost three times as large as the entire product spot market. The actual size may even be twice as large (26 mb/d) because the editors of *Argus* believe they are able to identify and confirm only 50 percent of the transactions.

The second principal forward market for crude oil is in Dubai crude. Equity production in this crude oil is shared between seven companies and production is fairly constant at 350,000 barrels/day. The standard cargo is 500,000 barrels. Trading started in late 1984 when spot trading in Saudi Arabian crude stopped. The volume of transactions on the Dubai market is reported to be between one-third and one-half the volume on the Brent market. Currently, the average volume of reported transactions would appear to be approximately 2 mb/d. However, many transactions in this market are also not reported. The editors of *Argus* believe that they are able to identify perhaps one transaction in three, suggesting that the total number of transactions may amount to 6 mb/d.

Other forward markets have developed for heating oil, other products, and other crude oils. The total volume of transactions in these markets probably ranges from 2 to 5 mb/d. Combined total volume of spot and forward transactions at the end of 1987 was approximately 45 million barrels per calendar day (63 million barrels per business day), slightly less than the 1987 free world consumption of 47 mb/d. The total value of transactions was roughly $700 million per calendar day.

The volume of trade on petroleum futures markets is approximately 50 percent greater than the volume on spot and futures markets. The statistics show that volumes involved in futures transactions amounted to more than 50 mb/d of crude oil, 13 mb/d of heating oil, and 8 mb/d of gasoline. In addition, trade in crude oil options exceeded 20 mb/d. Thus, average number of transactions in oil commodity markets exceeds 135 million barrels per calendar day (170 million barrels per business day). Trade in oil markets is more than double free world consumption. The rate of increase in volume has been 45 percent per year.

The Influence of Oil Commodity Markets. Petroleum commodity markets have become the mechanism through which the price of crude oil is determined. OPEC's posted price system was replaced by a formula that tied the buyer's payment to the publicly quoted price of products traded (later crude oil) in spot markets located near the ultimate destination in 1985 when netback pricing was introduced by Saudi Arabia in 1985. While the terms and philosophy underlying these arrangements have changed with the passage of time, the basic philosophy—that the seller must absorb most of the risk of price fluctuation—has remained constant. In fact, commodity-type pricing was so dominant that most traditional types of transactions no longer existed.

The dominance of commodity-type transactions is evident in data gathered by the editors of *Petroleum Intelligence Weekly*. The *PIW* data show that 66 percent of the transactions in the international crude market as of September 1988 were linked to commodity markets in one way or another while only 16 percent of the oil was purchased under the traditional posting system. The *PIW* report identified ten separate types of pricing arrangements. Six of these ten arrangements accounting for 66 percent of the exports of twenty-six countries were linked to commodity market–type instruments and could be referred to as "on arrival pricing." Another 10 percent of the volume was distributed through barter arrangements, where the price of the oil is linked to the spot market price.[15]

The greater use of commodity markets by firms in the industry reflects the need to establish protection against volatile prices by the ultimate consumers of crude oil—refiners. Refining margins have historically been small, ranging from $0.50 to $3 per barrel in the best of times and averaging only $0.37 per barrel. The *standard deviation* of crude oil prices, by contrast, has exceeded $3 per barrel since 1985. Under these circumstances refiners have been forced to turn to commodity markets to protect themselves from bankruptcy.

Conclusion

The 1986 oil price collapse has often been associated with the development of oil commodity markets. However, such an explanation overly simplifies the complicated process of structural change that occurred in oil over the last two decades.

As shown here, the collapse of oil prices occurred because concentration in the world oil market declined. The number of oil-exporting nations increased during the late 1970s and early 1980s while the market share of the dominant exporting nations declined. Concentration in the market fell, and a decline in prices followed. This decline should have been expected. Future changes in oil prices will depend upon trends in market concentration, not on the existence of surplus capacity, as asserted by some.

However, the process that led to the decline in market concentration was not entirely independent on the growth in oil commodity markets. Both changes resulted from the nationalization of the producing interests of the multinational companies. The accelerated development of reserves in countries that had not previously participated in the world oil market resulted from efforts by multinationals to replace reserves seized by OPEC. At the same time these companies came to rely on large volumes of oil purchased at arm's length. The growth of oil commodity markets was the natural consequence of this change.

Notes

This chapter is a preliminary version of Chapter 3 of *Preparing for the Next Energy Crisis*, forthcoming from the Institute for International Economics, Washington, D.C., 1991. Reprinted with permission of the Institute for International Economics.

1. M. A. Adelman, *The World Petroleum Market* (Washington, D.C.: Johns Hopkins University Press for Resources for the Future, 1972).

2. Ibid, p. 97.

3. Daniel Badger and Robert Belgrave, *Oil Supply and Price: What Went Right in 1980?* (British Institute's Joint Energy Policy Program, Policy Studies Institute, 1982), p. 120.

4. Adelman, *World Petroleum Market*, p. 156.

5. The "competitive fringe" is composed of producers that enjoy the benefits of cartel membership (higher prices) without incurring the obligations of membership. A less flattering name is "free riders." Ecbo provides a careful review of the actions of over fifty cartels. He notes that only a third of the attempts to cartelize a market have succeeded and finds that the average life span of a successful cartel is approximately four years. See Paul Ecbo, *The Future of World Oil* (Cambridge, Mass.: Ballinger Publishing, 1976).

6. Thomas Hieronymus, *Economics of Futures Trading*, 2nd ed. (New York: Commodity Research Bureau, 1977), p. 21.

7. Julius B. Baer and Olin Glen Saxon, *Commodity Exchange and Futures Trading* (New York: Harper and Bros., 1948), chap. 6.

8. Committee on Commodities Regulation of the Association of the Bar of the City of New York, "The Forward Contract Exclusion: An Analysis of Off-Exchange Commodity-Based Instruments," *Business Lawyer* 41 (May 1986): 854.

9. Jeffrey Williams suggests that this distinction is more imaginary than real, noting that the Chicago Board of Trade, a futures market, did not require the payment of margins and did not have a clearinghouse before the 1880s. Further, he notes that margins are still not paid between principals on the London Metal Exchange. Williams, *The Economic Function of Futures Markets* (Cambridge: Cambridge University Press, 1986), pp. 164–66.

10. Representatives of futures markets note that delivery terms can be altered through ADPs (alternative delivery procedures) or EFPs (exchanges of futures for physicals). However, these are specific delivery arrangements that are worked out between parties before or at the expiration of a contract. Without such mutual agreement, the terms of the contract must be fulfilled.

11. Joe Roeber Associates, *COMMA, the EEC Register of Spot Transactions, Summary and Conclusions of the Final Report*, (Commission of the European Economic Communities, Brussels, 22 October 1980), p. 11.

12. Robert Mabro et al., *The Market for North Sea Crude Oil* (Oxford: Oxford University Press, 1986), p. 21.

13. For a period from 1985 until August 1988 the lot size was increased to 600,000 barrels.

14. Mabro et al., *Market*, p. 163.

15. *Petroleum Intelligence Weekly*, 14 November 1988, p. 4.

CAROL A. DAHL

6. U.S. and World Oil Production and Production Costs in the 1980s

OIL markets left the 1970s like a lion, with shortages, high prices, and high costs creating confusion. The biggest drilling boom in U.S. history, which peaked in 1981 along with oil prices, managed to stem production and reserve declines, but the boom soon turned to bust. Prices began a descent that culminated in 1986, and exploration and development followed right behind. U.S. production, reserves, costs, and prices soon returned to the historical trends and patterns of the years prior to the mid-1970s. The lion seemed to have turned to a lamb as the decade of the eighties drew to a close.

Complacency has not entirely returned; memories of the lion and the more recent Iraqi invasion of Kuwait remain to motivate a closer look at this decade of turbulence. I will begin with the historical trends leading into the decade, including the relative position of the United States in world markets. Closer scrutiny of events in the 1980s will show how these trends were first reversed and then reinforced. Observing changes in production, reserves, exploration and development, costs, and their components will enhance understanding of the U.S. markets during the decade. Comparing these parameters across states will show where we can expect future U.S. production to occur. Comparing them internationally will hint at changes to come and show the United States as a high-cost declining reserve producer that can expect its role as a world supplier of oil to continue to diminish.

Historical Market Trends

In the 1960s and early 1970s energy supply was thought to be abundant and demand-driven. Oil was viewed as a swing fuel with an oil quota keeping out cheap foreign imports. Nuclear power was perhaps unlimited, and dirty old coal was on its way out. Remaining energy problems would succumb to a technological fix. Although Hubbert had forecast U.S. oil production to peak around 1970, most geologists were more optimistic.[1]

This mood changed as the decade of the 1970s progressed. As seen in Figures 6.1 and 6.2, the U.S. share of the world market, already

Figure 6.1. U.S. oil production, oil product demand, and oil reserves as of January 1 of each year.

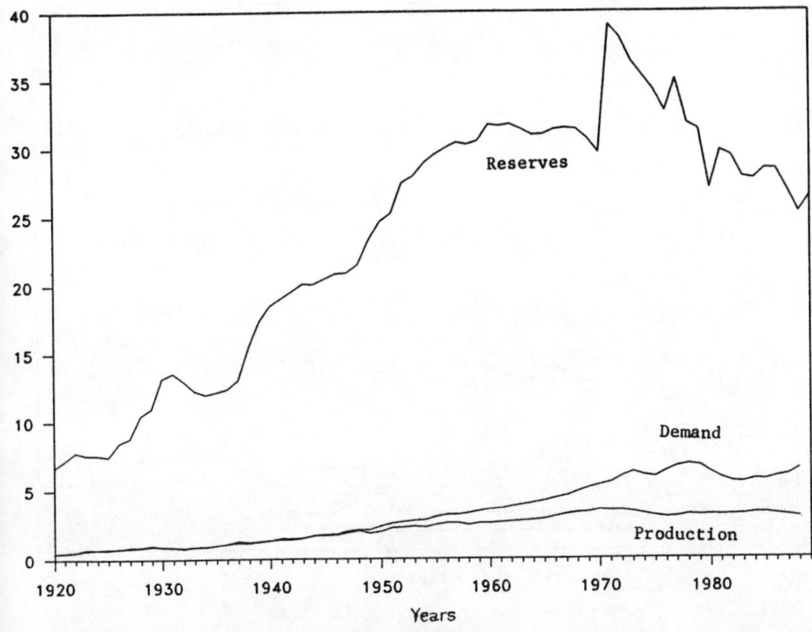

Sources: American Petroleum Institute, *Petroleum Facts and Figures* (1971), p. 115; American Petroleum Institute, *Basic Petroleum Data Book* (Janaury 1989), Section 2, Table 1, Section 4, Tables 1, 10, and Section 7, Table 1; U.S. Department of Commerce, *Historical Statistics of the United States Colonial Times to 1970* (1975), pp. 593–94; *Oil and Gas Journal,* Worldwide Report, 26 December 1988, p. 49; *Oil and Gas Journal,* Forecast Review Issue, 30 January 1989, p. 51

falling in relative terms, now fell in absolute terms as U.S. oil and gas production peaked early in the decade and fell by 1976 to its lowest level since 1966. The Club of Rome Report's *Limits to Growth* and increasing awareness of environmental degradation from nuclear power and coal suggested increasing energy scarcity in the years to come. The 1973–74 Arab oil embargo threatened security because of oil's strategic importance as a transportation fuel, its concentration in politically unstable areas, and a declining U.S. reserve base. In 1975 OPEC, with over three-fourths of total oil exports, was estimated to have 68 percent of the world's reserves and an even larger share of the exportable surplus, while the United States, with almost a third of world consumption, had less than 5 percent of total proven world reserves.

Beginning in 1976 U.S. oil production, including lease condensate, zigzagged its way up, increasing 5 percent between 1976 and 1980,

Figure 6.2. Oil reserves in the United States, Communist countries, OPEC, and the total world as of January 1 of each year.

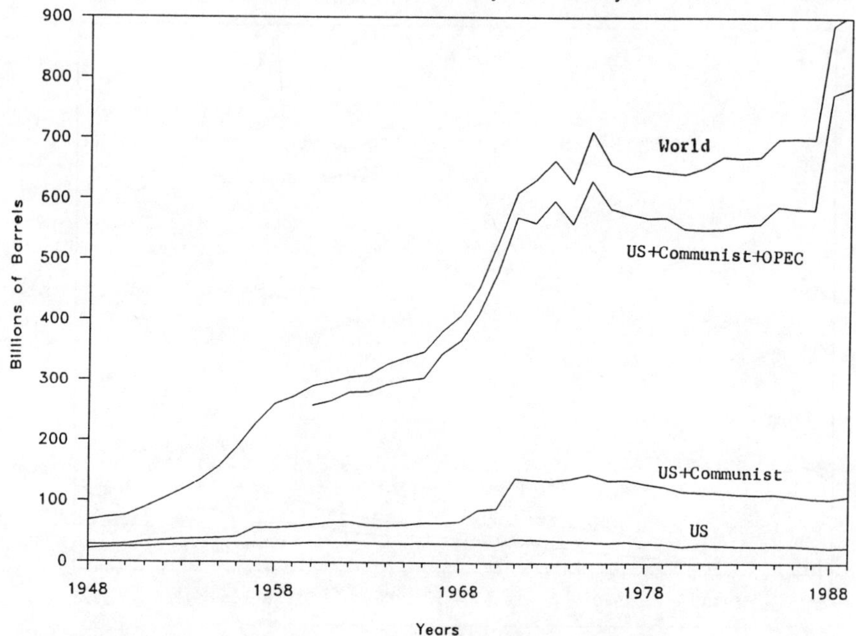

Sources: American Petroleum Institute, *Basic Petroleum Data Book* (January 1989), Section 2, Table 1; *Oil and Gas Journal,* Worldwide Report, 26 December 1988, pp. 48–49

dipping slightly in 1981, and increasing almost another 5 percent between 1981 and 1985. This trend was dramatically reversed with the oil price collapse. Oil production fell almost 15 percent between 1985 and 1989. World production patterns were somewhat different. Figure 6.3 shows world production peaking in 1979 but then falling from lack of market at the prevailing high prices with a dramatic shift away from OPEC to non-OPEC production after 1978. It also illustrates the changing share of the United States in oil markets and how small the shifts in U.S. production were relative to those worldwide.

U.S. oil product demand patterns, also seen in Figure 6.3, show a somewhat different pattern from U.S. oil production. Demand peaked in 1978, fell 20 percent from 1978 to 1983, but regained much of the decrease between 1983 and 1988. World product demand showed somewhat the same trend, peaking in 1979, falling 11 percent to 1983, and rising almost 10 percent between 1983 and 1988.

One component of increasing U.S. demand has been a 594 million barrel U.S. buildup of crude and products, seen in Figure 6.3. The

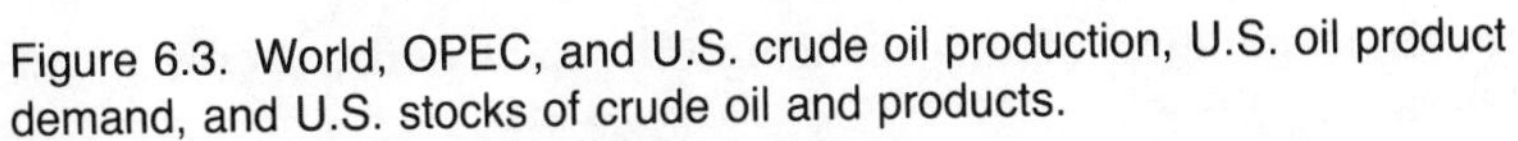

Figure 6.3. World, OPEC, and U.S. crude oil production, U.S. oil product demand, and U.S. stocks of crude oil and products.

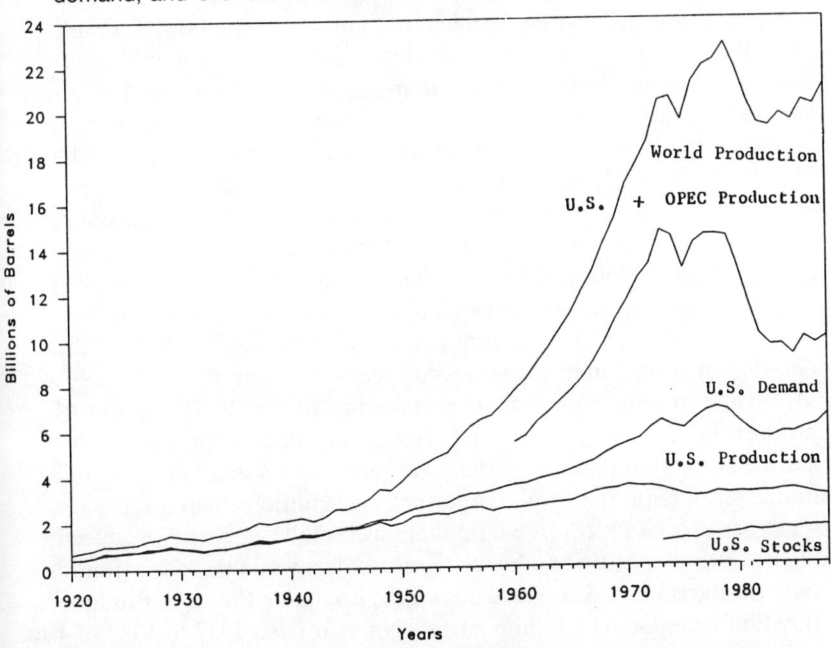

Sources: American Petroleum Institute, *Petroleum Facts and Figures* (1971), p. 548; American Petroleum Institute, *Basic Petroleum Data Book* (January 1989), Section 4, Table 1, Section 7, Table 1, and Section 14, Table 2; *Oil and Gas Journal,* Worldwide Report, 26 December 1988, pp. 48–49; *Oil and Gas Journal,* Forecast Review Issue, 30 January 1989, p. 51; U.S. Department of Energy-Energy Information Administration, *Monthly Energy Review* (October 1988): 119

bulk of this buildup comes from the strategic petroleum reserve, since private crude oil stocks are down somewhat from their 1982 high of almost 8 percent of product demand. Stocks for the OECD as a whole have been up as well with a buildup of 843 million barrels since 1973 but down somewhat since their 1980 high. Total product and crude oil stocks as a percent of product demand have varied by region with the percent of stocks remaining higher in western Europe and Canada than in the United States.

Oil Production in the 1980s

Economic theory suggests that production decisions depend on interest rates, oil prices, and the geological characteristics of oil reserves, which in turn determine the costs of finding, developing, and producing these reserves. A closer look at these variables will increase our understanding of the changing oil production patterns in the 1980s.

Interest rates have two possible effects on oil production. A simple Hotelling argument suggests that higher interest rates would make money in the bank worth more than oil in the ground. Producers should then raise production to hold money rather than oil. However, empirical work on the oil market has not been able to support the Hotelling type of behavior for the oil market. Alternatively, higher interest rates raise development costs, which should lower production. Since real interest rates were abnormally high in the 1980s, with high-grade corporate real bond yields averaging 6 percent from 1980 to 1986 compared to the previous postwar high of 1.1 percent, they would have detracted from production gains in the first half of the decade but contributed to the recent years of falling production.

Previous research finds that reserves are the best explanation of production in the postwar period. Hence an examination of reserves and reserve additions should help us better understand current and future production patterns. Larger reserve additions increase the reserve base. A larger reserve base that pushes the last barrel produced further into the future and hence renders it less valuable today should raise production. U.S. reserve additions, excluding the huge Prudhoe Bay find, averaged 3.1 billion barrels per year from 1947 to 1973, 2.4 billion barrels per year from 1974 to 1979, and 2.8 billion barrels per year from 1980 to 1988. The very low level of additions from 1974 to 1979 might be attributable to price controls that discriminated against enhancing reserves and production in known fields in favor of looking for new but less likely found oil. Although additions were up in 1980 to 1988 (probably driven by increased oil prices), they did not respond to large price increases as much as historical data would have suggested.[2]

Total proven reserves have tended to fall since 1970 as the United States has failed to find replacement reserves. The rate slowed after the mid-1970s, resumed in 1986 with the large dip in prices, and rebounded somewhat in 1987. In Figure 6.4 we can see the states listed separately that represent between 80 and 90 percent of U.S. proven reserves. The seemingly large losses in proven reserves in Louisiana and California in 1986 resulted from removing reserves in federal waters from the state category. The overall downward trend in U.S. reserves averaged 2.3 percent from 1970 to 1986 with the decline rates somewhat higher in Texas, Louisiana, and Oklahoma; almost at the U.S. average for Alaska; but lower in California, New Mexico, and Wyoming.

Although the reserve base falls with production, higher oil prices

Figure 6.4. Oil reserves as of January 1 of each year: Texas (TX), Alaska (AK), Louisiana (LA), California (CA), Oklahoma (OK), New Mexico (NM), Wyoming (WY), and the United States.

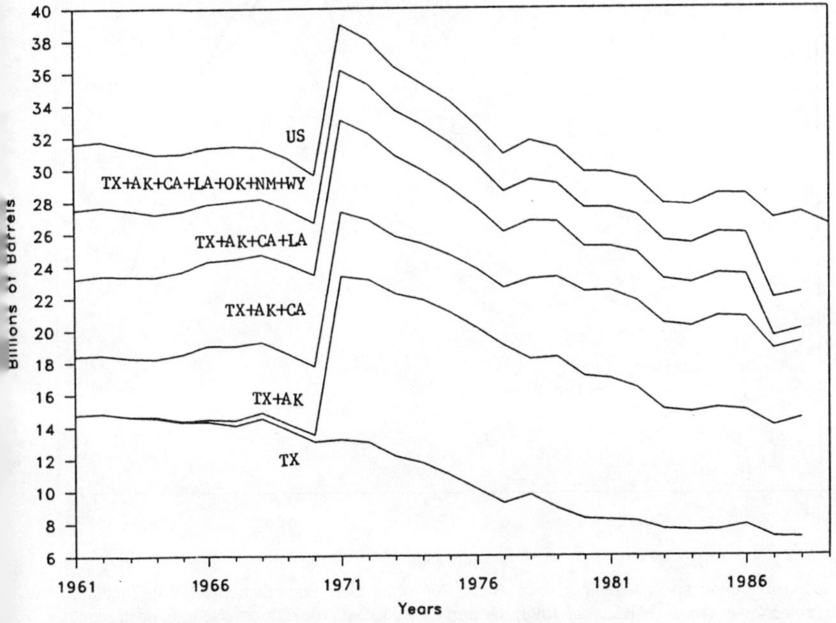

Sources: American Petroleum Institute, *Basic Petroleum Data Book* (January 1989), Section 2, Table 3; *World Oil,* Outlook Issue (February 1989): 76

should have made it more profitable to search for more reserves to replace those produced. Three possible reasons for not replacing reserves could be that we looked less hard, that reserves have been harder to find, or that we have been more inefficient at finding them. The exploration and development record suggests that although we have had almost as many drilling rigs and seismic crews as before, we have never drilled as many wells or as many feet as in the first half of the 1980s. Figure 6.5 shows the general pattern of exploration and development. Seismic crews and active rigs peaked in 1981 one year after real U.S. wellhead prices peaked and the windfall profit tax was passed. In the same year nominal U.S. wellhead prices peaked. Total rigs peaked a year later, showing a lag in adjustment, while the difference between active and total rigs illustrates the depression in the oil service sector discussed in Chapter 7.

Since wells and feet drilled show the same pattern as active rigs but with sharper fluctuations, the intensity of exploration and development efforts as measured by wells per rig and feet per rig in-

Figure 6.5. U.S. total and active drilling rigs and seismic crew months.

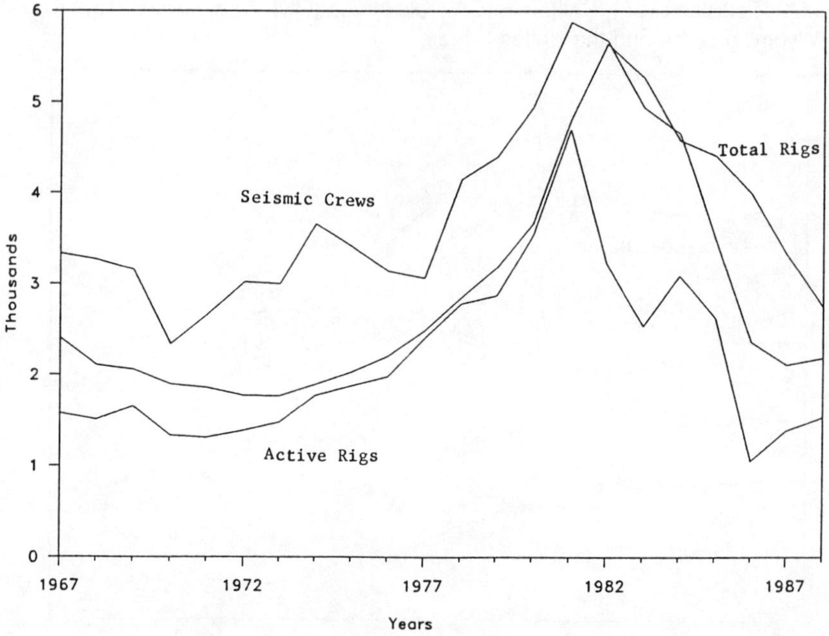

Sources: American Petroleum Institute, *Basic Petroleum Data Book* (January 1989), Section 3, Tables 4, 16; *World Oil* (October 1988); *Oil and Gas Journal* (1987), fourth issue of each month

creased when rig activity increased but decreased when rig activity decreased. Thus through 1985 we looked more and harder than ever before. However, since the percent of exploratory wells declined and the percent of new field wildcats declined even more, we were working relatively more intensely in old places. This reduction in exploratory drilling is reflected in a further decrease in the percent of dry holes in the 1980s, which had fairly steadily increased through the mid-1960s. Higher oil prices also decreased the dry hole ratio by making otherwise uneconomic wells profitable.

In addition to not looking as much in new places, we were not looking as deep. Well depth on average fell somewhat from the 1970s. New oil wells, in particular, are not appreciably deeper and are somewhat more shallow than in the 1970s.

Although drilling more, we were not replacing reserves. Reserves per well, per foot drilled, and per rig tended to be lower as discoveries of giant fields or fields with original oil in place in excess of 100 million barrels declined. In the United States in 1989 there are 227 giant fields still producing.[3] These fields, which constitute less than

1 percent of the over 23,000 U.S. oil fields, produced 60 percent of the U.S. oil in 1988, contained an estimated 65 percent of remaining proven reserves, and produced an average 28.3 barrels of oil per well per day compared to the U.S. average of 13.3. Figure 6.6 illustrates the importance of these fields to proven reserves with the top line being proven reserves in the given year, the bottom line barrels per well per day, and the histogram containing total reserves in giants, both produced and remaining as of 1988, plotted against year of discovery. Figure 6.6 also illustrates the lag between discovery of a field and the knowledge of its total reserves gained as the field is developed. For example, the extent of the huge reserves discovered in the 1930s was only realized in the following decades.

Even given the reduction in giant fields found (only 13 of the 227 were found since 1973 and only 4 of the 227 have been found since 1980), total reserve additions in the 1980s have exceeded those in the 1960s and those in the 1970s as well if we exclude Prudhoe Bay. However, these additions have come with a great deal of drilling

Figure 6.6. U.S. proven oil reserves as of January 1, ultimate reserves in giants as of 1989 by discovery date.

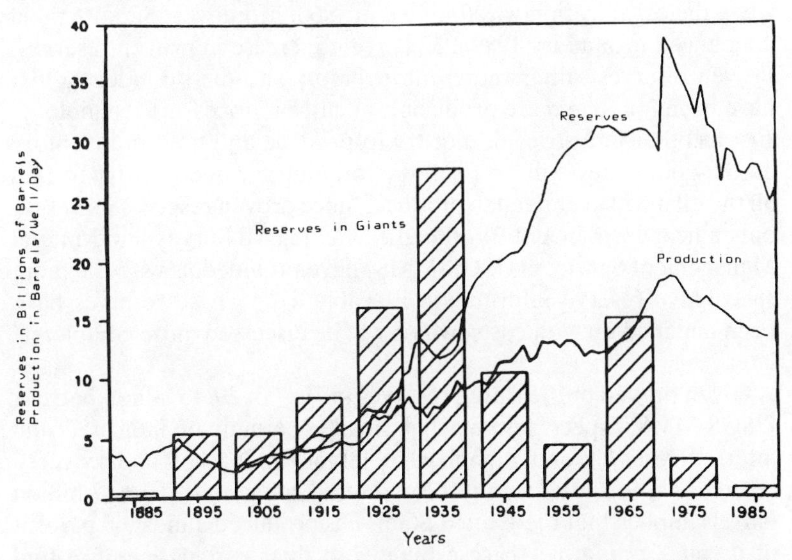

Sources: American Petroleum Institute, *Petroleum Facts and Figures* (1971), pp. 115, 548; American Petroleum Institute, *Basic Petroleum Data Book* (January 1989), Section 2, Table 1, Section 3, Table 18, and Section 4, Table 1; U.S. Department of Commerce, *Historical Statistics of the United States Colonial times to 1970* (1975), pp. 593–94; *Oil and Gas Journal,* Worldwide Report, 26 December 1988, p. 49; *Oil and Gas Journal,* Forecast Review Issue, 30 January 1989, pp. 69–70; World Oil, *Outlook Issue* (February 1989), 64

effort, and production per well has continued the descent that began in 1970.

Although reserve additions were up in the early 1980s they have not increased as fast as expected given the extraordinary amount of exploratory effort. With the immense buildup in rigs between 1973 and 1982 and little excess capacity until after 1981, efficiency of exploration measured as reserves per foot, per well, and per active rig tended to be inversely related to the level of exploration. With the big spurt in new rigs between 1978 and 1982, finding rates per active rig were exceptionally low, probably the result of highly speculative drilling coupled with less experienced operators. The active rig count has continued to fall, leaving a great deal of excess capacity, a highly depressed drilling industry, but the most productive drillers. Reserves per rig, per well, and per foot have continued to rise. Thus, drilling efficiency has gone up. However, since feet and wells per active rig have fallen considerably from their 1981 peak, drilling intensity has fallen.

Proven reserves do not represent all the oil that exists. For example, over five times as many additional reserves have been found in the United States as were proven in 1947. Over 1.5 times as many additional reserves have been found in OPEC countries as were proven when the organization was founded in 1960. Moreover, Saudi Arabia announced in January 1989 a 52 percent increase in proven reserves. Proven reserves rather represent inventories to the oil industry that have been found and are producible at current prices and technology. Just as other industries do not try to produce and hold in inventory all the goods they will ever sell, the oil industry does not try to find all the oil it will ever need to produce. Since proven reserves represent only a near-term inventory, the U.S. Geological Survey and Mineral Management Services (USGS-MMS) have provided us with the more speculative reserve information in Table 6.1.[4] These reserves have been matched up with costs, which will be discussed more completely later.[5]

Given proved or measured reserves in 1987 of 29.46 billion barrels, USGS-MMS expect a somewhat smaller amount of indicated and inferred reserves and a somewhat larger amount of undiscovered reserves.[6] Their total for all three reserve categories of 100.7 billion barrels implies that the United States has produced almost 60 percent of its oil. Comparing these estimates to their earlier ones, we find that they have become more pessimistic in their outlook for the total United States, offshore Atlantic, and the Rocky Mountain and Colorado regions but more optimistic for the U.S. Gulf.[7]

From Table 6.1 we can see current production patterns and spec-

6.1. U.S. Oil: Costs, Production, and Reserves, 1986–1987 (1987 dollars per barrel and ns of barrels)

	Cost in U.S. dollars	Production	Cumulative	Reserves			
al offshore							
a offshore	—	—	—	—	—	3.4	3.4
c offshore	$4.14	0.031	0.4	1.3	0.2	3.4	4.9
ffshore	3.87	0.326	6.9	4.0	0.4	8.6	13.0
c offshore	—	—	—	—	—	0.7	0.7
onshore and offshore							
a	1.31	0.717	6.1	6.9	6.4	13.2	26.5
	2.74	0.365	20.3	4.7	1.2	3.5	9.4
do Plateau and Basin Mountain Range	4.02	0.132	3.1	0.6	0.4	1.5	2.5
Mountains and Northern at Plains	14.89	0.148	7.1	1.2	1.3	4.5	7.0
Texas	11.36	0.520	30.2	5.4	3.8	2.6	11.8
	9.53	0.486	43.1	3.7	5.7	4.2	13.6
tinent	13.92	0.194	17.3	1.1	1.4	1.9	4.4
n Interior	13.06	0.053	8.4	0.5	0.7	1.8	3.0
c Coast	10.27	0.830	0.1	0.0	0.0	0.2	0.2
tal	6.58	3.801	142.9	29.5	21.7	49.4	100.7

finitions and sources: Column 1 is in-ground exploration and development costs in 1987 dollars. They are computed gregated by the author using U.S. Energy Administration, *Crude Oil, Natural Gas and Natural Gas Liquids Reserves* pp. 75–84; *The Joint Survey of Drilling* (1986–1987), Table 1: and *World Oil*, Forecast Review Issues (February 989). Column 2 is 1987 production aggregated by the author, from *World Oil*, Forecast Review Issue (February 1989). 3 is cumulative production to 1986. Columns 4–6 are from the U.S. Geological Survey and Mineral Management *Estimates of Undiscovered Conventional Oil and Gas Resources in the United States—A Part of the Nation's Endowment* (1989). Column 4 is measured reserves. Column 5 is inferred and indicated reserves. Column 6 is vered reserves. Column 7 is the sum of columns 4, 5, and 6. Columns 2–7 are measured in billions of barrels (bb).

ulate how U.S. production patterns will change. Alaska, a younger producing province, is currently producing 22 percent of U.S. crude oil with roughly the same percentage of measured reserves. Offshore production, which is 14 percent of the U.S. total but 20 percent of the known reserves, is being exploited less rapidly than older onshore reserves. Alaska and federal offshore may each contain more than a fourth of inferred, indicated, and undiscovered reserves, implying that they may eventually provide over half of U.S. production. The only other area where the percent of reserves exceeds the current percent of production, suggesting an increasing share, is the Rocky Mountain region.

Even though reserve additions in most years have not kept up, production tended to be up in the early 1980s. Figure 6.7 shows the

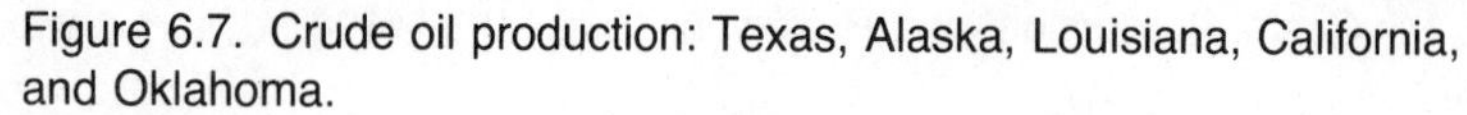

Figure 6.7. Crude oil production: Texas, Alaska, Louisiana, California, and Oklahoma.

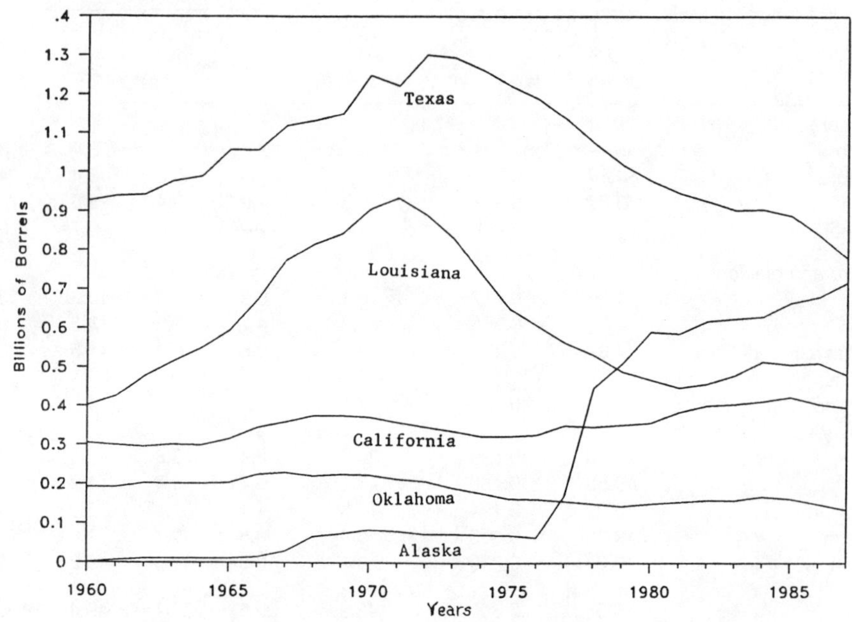

Sources: American Petroleum Institute, *Basic Petroleum Data Book* (January 1989), Section 4, Table 4; *World Oil*, Outlook Issue (February 1989): 62

production profile of the five largest states which produce over 80 percent of U.S. crude. Texas has shown an almost continual production decline since the early 1970s despite continual increases in drilling. Particularly large amounts of drilling from 1978 to 1985 only managed to stem the fall from 1983 to 1985. Oklahoma, with a fairly large number of well completions, managed only a slight increase in production before the decline resumed. Production in Louisiana and California tended to increase through 1985 before declining again. Of the major producers only Alaska, with production rising over the entire period, has offset some of these declines, while Kansas, Utah, and North Dakota account for much of the rest of the increase.

Production gains were made in another relatively new area, federal offshore, and by working old reserves more intensely. For example, operating existing wells longer increases stripper production,[8] and enhanced recovery techniques can increase the amount of oil recovered from existing fields now estimated to be 40 percent.[9] Figure 6.8 shows the production profiles of these additional categories to help further isolate the origin of the 146 million barrel increase in

Figure 6.8. Enhanced oil recovery, offshore and stripper well production, 1980–1988.

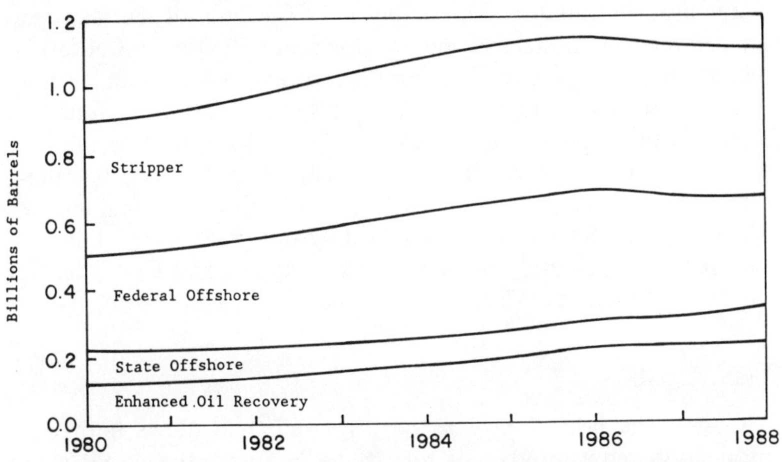

Sources: Offshore and stripper production taken from the American Petroleum Institute (1989), *Basic Petroleum Databook,* January, Section IV, Table 5 and Section XI, Table 18, and U.S. EIA/DOE (1988), *Petroleum Supply Annual,* January, Volume 1, Table 9. Enhanced oil recovery taken from *Oil and Gas Journal* (1988), Annual Production Report, 18 April 1988, p. 46. The bottom line is EOR, next line is EOR and state offshore, etc.

total U.S. production from 1981 to 1985. I will proceed by discussing in somewhat more detail their production profiles on a regional basis.

Part of the overall increase and most of the increase in Oklahoma through 1985 came from stripper production. The 17 percent total U.S. increase was roughtly the same as the increase in the number of stripper wells and was accompanied by an increase in the percent of crude from stripper wells. Somewhat over half the increase in wells appears to be from reduced abandonments, the rest from other wells' production falling enough to put them into the stripper category. Stripper abandonments were lowest in 1981 but continued to increase, exceeding the pre-1973 level by 1987.

Another source of increased production came from federal offshore, while state offshore, except in Alaska, continued to decline. The bulk of the increases were in Louisiana with most of the rest in California. Much of Louisiana's increase came from this offshore production, which more than offset the decline from Louisiana onshore. Half of California's production increase came from increasing stripper and federal offshore production with the rest from onshore.

Last, enhanced oil recovery (EOR) has also been a source of increased production from existing reservoirs. There are three important types of EOR. The most important, accounting for almost three-fourths by volume, is thermal recovery, or applying heat to heavy crudes to decrease their viscosity. It includes steam flooding, hot water injection, and in situ combustion. The bulk of thermally enhanced recovery is steam flooding. It is primarily used in California, where enhanced recovery onshore more than offset other onshore declines. Second in importance, accounting for almost one-fourth of EOR, is injection of such gases as CO_2 and nitrogen to increase pressure and force more oil to the surface. Chemical projects trailing a far third account for less than 5 percent of EOR. The total number of EOR projects fell rather dramatically from 1986 to 1988, but thermal recovery volumes fell only a small amount and CO_2 injection increased enough to offset these declines.

U.S. Costs

Cost is the last variable to be discussed that affects both U.S. production and its production relative to the rest of the world. There are three categories of costs for producing a barrel of oil. The cost of exploring or finding oil reserves includes drilling and equipping wells, acquiring acreage, geological and geophysical work, and lease rental. The cost of developing reserves or putting in oil wells and other necessary infrastructure includes drilling and equipping wells, equipment lease, acquiring producing acreage, and improved recovery programs. Since the distinction between the exploration and development categories is blurred, they will be combined. Once the infrastructure is in place, production costs associated with each barrel lifted include taxes and direct lifting costs such as salt water disposal, fuel, and royalties.

Everything else equal, unit cost should rise over time as we deplete reservoirs and reserves become increasingly difficult to find, but they should fall with improved technology. Figure 6.9 shows estimates of how costs have changed over the postwar era for the average U.S. barrel. Since they are average U.S. costs of oil in the ground, they do not include the normal rate of return that an oil company would require to produce the oil.[10]

Costs were relatively flat from 1947 to 1972 except for the dip associated with the 1968 Prudhoe Bay find. Costs followed oil prices up, with exploration and development costs peaking in 1982 and production costs peaking in 1981. They then followed prices back down. The big drop in costs in 1987 resulted partially from large reserve additions, which corrected for the pessimistic downward revisions in 1986. Hence, the average of 1986 and 1987 of $6.39 per barrel may be more comparable to earlier numbers. This cost is about

Figure 6.9. U.S. wellhead price of crude oil, total cost of producing crude oil, and cost of exploration and development of crude oil.

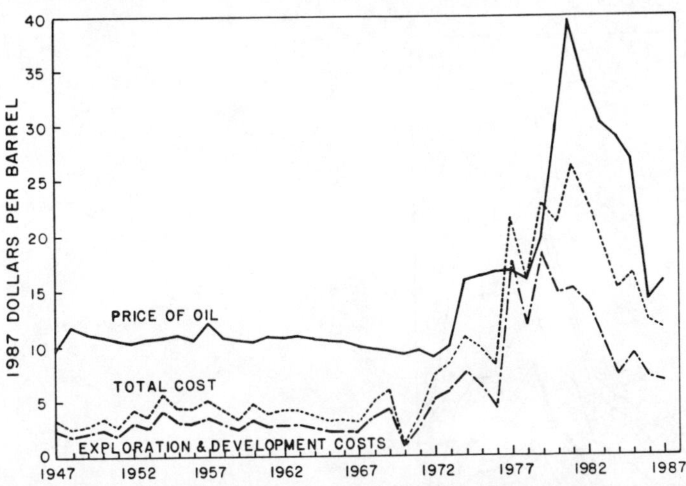

Sources: Exploration and development (finding costs) are computed from total exploration and development expenditures from Chase Econometrics for 1947–84 and from the American Petroleum Institute for 1985–87 both in the *Basic Petoleum Databook* (January 1989) Section 5, Tables 9 and 10. These costs are distributed over new oil and gas reserves using weights computed from typical oil and gas wells given in *Oil and Gas Journal,* 20 February 1989, p. 49. Production (lifting costs) for 1947–72 are assumed to be the same percentage of exploration and development costs as in 1973. Production costs for 1973–77 are computed from production expenditures from the U.S. Department of Commerce given in the American Petroleum Institute, *Basic Petroleum Databook* (January 1989), Section 5, Table 10, distributing costs over oil and gas production using weights computed from typical oil and gas wells given in *Oil and Gas Journal* (20 February 1989), p. 49. Production costs for 1978–1987 are taken from U.S. EIA/DOE *Performance Profile of Major Energy Producers* (1984) and (1987)

equal to costs computed for 1973 but is roughly double the pre-1970 values.[11]

As oil prices went up, rents or excess profits were generated. To determine whether the increases in cost were real physical increases or were shares in these rents, oil company expenditures or costs are broken into their components for examination in Figure 6.10. We can see that from 1978 production expenditures—lifting costs and taxes—escalated the most and remained the largest share through 1986.

Total expenditures peaked in 1981 before declining. A further breakdown shows that taxes were the largest component at 44 percent of the total increase, particularly the windfall profits tax. This tax, effective from March 1980, became ineffective below about $20 per barrel and was repealed in August 1988. Much of the decrease in production costs per barrel from their 1981 peak resulted from tax

Figure 6.10. Oil company expenditures by category (millions of 1987 U.S. dollars).

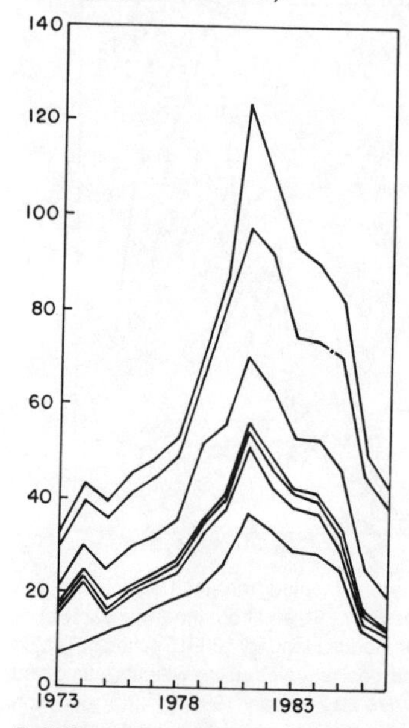

Definitions: Lines from top to bottom are expenditures for: 1, drilling and equipping wells; 2, 1+ acreage; 3, 2+ geophysical; 4, 3+ improved recovery; 5, 4+ total exploration and development; 6, 5+ lifting; and 7, 6+ taxes and is equivalent to total expenditures

Source: American Petroleum Institute, *Basic Petroleum Databook* (January 1989), Section 5, Table 10

decreases, since direct lifting costs per barrel tended to increase through 1985.

A significant portion of nontax lifting costs is salt water disposal, with around five barrels of water produced in the United States for every barrel of oil.[12] As regulations have tightened there has been less surface and marine disposal of this water, requiring reinjection wells. Aggregate pollution control costs for the oil and gas industry show a steady increase from $.21 per barrel of petroleum product demand in 1966 to $1.01 in 1982 but falling to $.75 per barrel by 1984.[13] With the recent series of accidents, particularly the Exxon *Valdez* running aground, these expenditures are expected to increase.

More low-cost gas offshore and retirement of uneconomic wells lowered direct lifting costs per barrel in 1986 after trending up for nearly a decade. Production costs onshore, totaling $5.13 including taxes in 1987, have generally been lower than production costs offshore, which were $2.93 including taxes in that same year. Less EOR as well as less expensive EOR contributed to falling direct lifting costs in 1986. EOR costs peaked at $14.20 per barrel in 1982 but fell to an average of $5.16 per barrel by 1986.[14]

Drilling, the second largest expenditure component, accounts for a third of the cost run-up. Since drilling expenditures increased 57 percent while drilling costs per foot increased 22 percent, somewhat under half the drilling expenditure run-up is accounted for by increased drilling costs per foot. These increased costs cannot be explained by deeper wells since the average well was only marginally deeper; neither are they the result of increased offshore drilling since onshore footage increased over twice as fast as that offshore.[15] Costs per foot by 1987 had returned to their 1973 level, suggesting that the cost run-up resulted from shortages and drilling rigs being able to garnish some of the rents accruing from the large price increases.

Total geophysical exploration costs accounted for 9 percent of the total cost run-up. Since costs went up over twice as fast as crew months, over half the increase resulted from an increase in unit cost. However, costs per crew month kept increasing after 1981 while the crew count fell almost half, suggesting a real increase rather than just increasing rents. This rather puzzling result, which could be a reflection of changing composition toward improved higher cost technology, would warrant further investigation.

Lease acquisition cost, which accounted for another 9 percent of the run-up, resulted from increased acreage and higher lease prices. The only leasing acreage data that are readily available are for offshore leasing, which accounted for more than half of acreage acquisition costs in 1978.[16] For offshore more than half the price run-up from 1978 to 1981 was an increase in leasing costs per acre. Price per acre for federal offshore acreage has fallen dramatically from a 1980 peak while leasing has tended to remain high except for the dramatic drop in 1986. Federal offshore acreage offerings have also remained exceptionally high, averaging over 95 million acres per year from 1983 to 1987 compared to the less than 5 million acre per year average of the previous decade.

The above discussion shows that up to 50 percent of the price increase was rents accruing to the government through increased taxes and lease acquisition costs. Kemp and Rose in an international comparison of taxing regimes for sixteen different fiscal systems came to a similar conclusion that for low- to medium-cost lower forty-eight fields the government's share of rents was close to 50 percent but was even higher for high-cost U.S. fields.[17]

The above costs have been derived using aggregate U.S. expenditure and are costs for the average U.S. barrel. However, costs are by no means uniform across states, while the state variation has implications for future markets. Regional expenditure data are not available to do a similar computation, but using drilling data and reserve additions by region I have computed unit costs for exploration and development for 1980 and 1986–87 by fairly detailed regions.[18] The average of 1986 and 1987 is used because the large reserve additions in 1987 are thought to be a correction for the pessimistic additions in 1986. A summary of my results aggregated by broad region are included in Table 6.1.[19] These again are costs of oil in the ground and do not contain a normal rate of return.

U.S. oil finding and development costs were $6.58 per barrel in 1986–87, down from $10.83 in 1980 with a rather wide variation in costs across regions from an Alaskan low of $1.31 to highs of over $25 per barrel in state offshore, Appalachia, and a couple of Texas districts. Low costs and relatively high reserves in Alaska pull average U.S. costs down, leaving them at approximately two-thirds those of the lower forty-eight. In general the West has cheaper reserves than the Midwest, South, and East, and newer provinces have cheaper reserves than older ones.

The average U.S. cost decrease between 1980 and 1986–87 was about 40 percent, with the percentage decrease in the lower forty-eight somewhat lower. Although costs rose somewhat for Texas they fell or were fairly flat in many major producing regions. Alaska has very low finding costs despite the fact that oil wells there tend to be twice as deep as the U.S. average and more than ten times as expensive to drill. These low costs, however, are offset by transport costs to the West Coast mainland of over $4 per barrel and a heavier, less valuable crude oil.[20] Federal offshore drilling tends to be even more expensive with wells again over twice as deep and even more expensive than those in Alaska. Yet federal offshore exploration and development with large reserve additions cost less than $5 per barrel, which is less than the national average and far less than state offshore, which cost over $25 per barrel to find and develop in 1986–87.

The distribution of costs in the United States can be seen in Figure 6.11, where costs are ordered and plotted against cumulative proven reserves. Over 80 percent of U.S. proven reserves lies in areas with 1986–87 finding and development costs of less than $6 per barrel, and over 90 percent lies in areas with 1986–87 finding and devel-

Figure 6.11. U.S. exploration and development cost per barrel, 1986–1987.

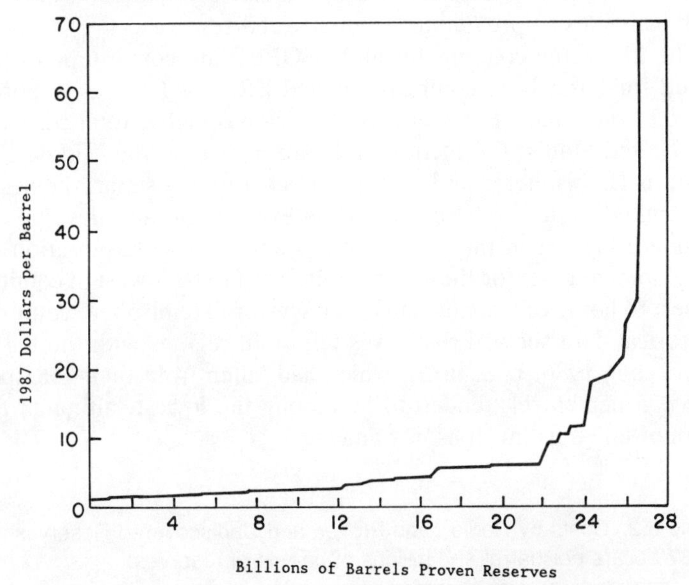

Sources: Author's computations using the U.S. EIA/DOE (1988), *Crude Oil, Natural Gas and Natural Gas Reserves,* pp. 75–84; *The Joint Survey of Drilling* (1986, 1987), Table 1; and *World Oil,* Forecast Review Issues, (February 1988, 1989)

opment costs of less than $15 per barrel. Since the reserves that are more expensive to produce have been developed, they raise the average cost of existing reserves. By matching these costs with the USGS's mean estimates of inferred, indicated, and undiscovered reserves in Table 6.1, we get an estimate of future costs in these areas. If recent cost decreases have come to an end, they might provide a reasonable lower bound for new costs in these regions. The U.S. average of $6.58 might provide a lower bound for future average U.S. costs.

U.S. and International Costs Compared

International costs are harder to acquire since they tend to be proprietary with public access more anecdotal and sporadic in nature. I have compiled data from the three main sources that are readily available: costs by broad region for large American companies operating in foreign countries computed by the U.S. Energy Information Administration, costs computed in the North Sea by fields, and costs computed using U.S. cost information and foreign drilling, well, and production data.[21] Selected cost estimates are given in Table 6.2 along with estimates of proven and undiscovered reserves.

The first three columns for all but OPEC are cost estimates published for large U.S. companies called FRS, or Federal Reporting Service companies. For them only Africa has higher total costs than the United States. Production taxes are highest in the Middle East, Eastern Hemisphere, and Africa, direct lifting costs are highest in the United States and Europe, while exploration and development costs are highest in the United States and Africa. Exploration and development costs for these companies tend to be lowest in countries closest to home or Canada and other Western Hemisphere countries. Historical data showed that taxes fell in all regions with the big dip in prices in 1986. U.S. taxes, which had fallen from their 1981 peak of $5.75 per barrel, tended to be among the lowest, although they had not fallen as much as in Canada.

Table 6.2. Costs by Region and Proven and Undiscovered Reserves (1987 dollars per barrel and billions of barrels of reserves)

	American Companies			*Reserves 1989*	
	Lift	*Tax*	*Exploration and Development*	*Proven*	*Undiscovered*
OPEC	—	—	$0.17	670.96	49.04
United States	$4.04	$0.57	6.76	26.50	67.90
Canada	3.60	0.42	2.93	6.79	32.16
Europe	4.07	0.80	3.61	28.98	32.62
Africa	2.67	2.44	10.29	56.44	67.06
Middle East	2.53	2.66	4.77	562.08	—
Other Eastern Hemisphere	2.17	2.62	5.77	44.66	46.59
Other Western Hemisphere	2.43	1.78	2.91	121.84	58.26

Data definitions and sources: Columns 1, 2, and 3 for all but OPEC are lifting costs, tax and exploration and development costs per barrel, respectively, computed from expenditure and reserve additions for large American companies operating abroad by the U.S. Energy Information Administration, *Performance Profiles of Major Energy Producers* (1987); Column 3 for OPEC is computed by the author using American drilling costs and foreign drilling, well and production information. Columns 4 and 5 are proven and undiscovered reserves in billions of barrels. Proven reserves are taken from *Oil and Gas Journal*, Worldwide Report, 26 December 1988, pp. 48–49. Proven reserves for other Eastern Hemisphere include the People's Republic of China but undiscovered reserves do not. Undiscovered reserves are computed by subtracting proven reserves in 1989 from the average of the 5th and 95th percentile estimate of total resources for 1985 taken from the U.S. Geological Survey, *World Resources of Crude Oil, Natural Gas, Natural Bitumen, and Shale Oil* (1987), prepared for the 11th World Petroleum Congress. Because Middle Eastern proven reserves had already exceeded the USGS total resource estimate by 1989, the undiscovered reserves column is left blank.

From more detailed operating and development costs, which exist on a field by field basis for the North Sea, I found the weighted average of operating costs is $3.85, which is fairly close to total lifting costs for American companies for Organization for Economic Co-operation and Development Europe. Eighty-two percent of remaining reserves lies in fields with current lifting costs below $5 per barrel. The weighted average of development costs in the ground is $4.62 per barrel, with 82 percent of reserves lying in fields with development costs less than $10. These development costs are somewhat higher than the finding costs for U.S. companies of $3.61. A further 2.45 billion barrels of Norwegian oil could be developed at in-the-ground costs of between $3 and $7 per barrel.

Since over 80 percent of OPEC crude oil is now participation crude or is controlled by the OPEC countries themselves, the above costs are not representative of OPEC production costs. However, OPEC costs are not publicly available and reserve statistics are not considered to be particularly reliable. Hence, in-ground OPEC costs in Table 6.2 are arrived at indirectly by using U.S. drilling costs but wells drilled and average production for individual countries, both considered to be more reliable than reserve numbers.[22]

The cost numbers computed by country tend to be lower than costs reported for large American companies operating overseas, except for Canada. Hence, they suggest even more strongly that the United States is a high-cost producer by world standards. They imply that OPEC has exceptionally low costs, that almost 80 percent of the world reserves lies in areas where finding and development costs excluding a normal rate of return may be less than $1 per barrel, and that 60 percent of world reserves lies in very low cost areas in the Middle East.

Because companies are not likely to have an incentive to understate their expenses or overstate their reserve additions, we might not expect the costs computed from company data to be overstated. Further, to the extent that American companies are less concentrated in the lowest-cost areas because many OPEC countries have taken over much of the responsibility for their own production and development, there may be an upper limit on costs for the region as well. On the other hand, needs for infrastructure, problems in dealing with unfamiliar cultures, and extra transport to import equipment and supplies may raise both nondrilling and drilling costs above those in the United States, implying that the numbers computed using U.S. costs, although perhaps technically feasible, may be lower on actual costs.

No matter which set of numbers seems closer to capturing reality, the United States still comes out a relatively high-cost producer. These differences in costs between the United States and foreign oil have caused investment patterns to change in the 1980s. Except for the 1984 sharp increase in development expenditures in Africa, Canada, and the United States, expenditures have trended downward in the United States while remaining more stable overseas, causing a relative shift toward cheaper foreign areas. Large U.S. companies have the largest percent of their foreign oil and gas production (37.5 percent of their foreign oil and 46.5 percent of their foreign gas) in OECD European countries. Their exploration and development expenditures have remained the highest there. Those in the Middle East strongly dominated by OPEC oil producers have remained among the lowest.

When these costs in Table 6.2 are matched up with proven reserves and undiscovered reserves, they rather dramatically illustrate the dominance that we might expect from OPEC in the years to come. Further, since mean undiscovered reserves forecast for the Middle East have already been exceeded, Table 6.2 understates the possibilities for OPEC.

Implications for the Future

U.S. dominance of world oil markets has been waning. With a great deal of effort in the early 1980s this trend was slowed as the U.S. share of world production increased. Production from Alaska, federal offshore, enhanced oil recovery, and lower stripper abandonments all served to increase oil production through 1985. The biggest drilling boom in history managed to slow the reserve decline, but not very efficiently, as reserves per foot and per drilling rig declined. Finds of giant fields which had fueled earlier reserve additions declined.

However, while the United States struggled to find reserves and continued to produce more than it found, reserves in the rest of the world continued to increase, particularly those in the Middle East. The U.S. reserve to production ratio is currently less than nine years, with inferred, indicated, and estimated mean undiscovered reserves adding another sixteen years. However, the global oil picture appears much brighter. For market economies the reserve over production ratio is over fifty years, with an estimated mean undiscovered reserves over production ratio adding another seventeen years.

A more detailed look at production, cost, and reserve additions in this chapter supports the contention elsewhere in this volume that the United States, particularly the lower forty-eight, is currently a high-cost area. Recent average in-ground exploration and development costs in the United States exceeded $6 per barrel, while elsewhere up to 60 percent of world reserves may have exploration and development costs under $1 per barrel.

The United States, the most highly drilled area of the world, may have a few surprises left. New technologies such as horizontal drilling or drilling at an angle to increase the producing surface and computer reservoir simulation as well as others yet to be discovered will help out. Studying the production patterns and reserve additions of the oldest petroleum provinces may provide clues as to what we can expect as younger regions mature. Newer U.S. areas such as Alaska and federal offshore appear more promising, but they have not recently had enough production to offset the declines in the older high-cost fields. Offshore Atlantic has been disappointing to date. Texas, once the largest U.S. producer, has had continued production declines and cost increases. If these trends continue they suggest that exploration and production are likely to continue their gradual move elsewhere. Barring policies or events that raise the costs of foreign oil, decrease the cost of domestic oil, or raise the cost of petroleum products to choke off U.S. demand, we can expect imports of crude oil and products inching toward 50 percent to continue their gradual increase.

Although the price collapse of 1986 is what sticks in our minds, exploration and reserve indicators had already been on their way down from their peaks in the early 1980s; the collapse merely accelerated things. At least half the large cost increases of the early 1980s were the result of increases in taxes. These and other cost increases had largely dissipated by the start of the 1990s; prices were much closer to historical averages. That violent market fluctuations have yielded to historical trends suggests that crisis-driven prices can reverse the trends, but not for long. Since the prudent producer, facing the first oil crisis of the 1990s, has this lesson of the 1980s well in mind, the deviation from trend may be lower and the postcrisis return to trend more rapid.

Notes

1. M. King Hubbert, *Nuclear Energy and the Fossil Fuels* (Dallas: American Petroleum Institute, Drilling and Production Practice, 1956), pp. 7–25, and M. King Hubbert, "Degree of Advancement of Petroleum Exploration in the U.S.," *American Association of Petroleum Geologists Bulletin* 51, no. 11 (1967): 2207–27.

2. See Carol A. Dahl, "Worldwide Comparison of Petroleum Production and Costs," Working Paper, Oxford Energy Institute, Oxford University, Oxford, England.

3. These fields, along with production, reserve, and well information,

are given each year in *Oil and Gas Journal*, Forecast Review Issue, published the last week in January. For a more complete discussion of giant fields, see *Oil and Gas Journal*, 16 December 1988, pp. 139–41, and *Oil and Gas Journal*, 2 January 1989, pp. 52–55.

4. U.S. Department of the Interior, *Estimates of Undiscovered Conventional Oil and Gas Resources in the United States—Part of the Nation's Energy Endowment* (Washington, D.C.: U.S. Geological Survey and Mineral Management Services, 1989).

5. Dahl, "Worldwide Comparison."

6. Basically, measured reserves are those we can expect to produce given current prices, technology, and geological knowledge. Indicated reserves are those we expect to be able to produce using improved recovery techniques such as fluid injection. Inferred reserves are those that we expect to add through extensions, revisions, and new pay zones. Undiscovered resources are those outside of known fields but expected to be found given broad geologic knowledge and theory.

7. U.S. Department of the Interior, "Estimates of Undiscovered Recoverable Conventional Resources of Oil and Gas," U.S. Geological Survey Circular 860, Washington, D.C., 1981.

8. Stripper wells produce fewer than ten barrels per day, but given their preferential tax treatment their exact interpretation is not altogether clear.

9. John D. Moody and Robert E. Geiger, "Petroleum Reserves: How Much Oil and Where?" *Technology Review* (March/April 1975): 40.

10. One can easily approximate an above-ground cost or an oil price (P) necessary to cover these costs or per barrel exploration and development (Ce & d) and production costs including taxes (Cp) plus a normal rate of return (r) by the formula P = Ce & d (r + d)/d + Cp (1 + r), where d is the depletion rate of a field typically around .1 for the United States. If r is the real rate of return, then P is the real price of oil; if r is the nominal rate of return, which includes the rate of inflation, then P is the nominal price of oil. Dahl, "Worldwide Comparison."

11. These costs are computed using reserve additions in the year that the expenditures accrue. There is, however, a lag between reserve additions and some expenditures. If a distributed lag on expenditures is used the cost curve is smoothed and moved to the right.

12. Joseph R. Dancy and Victoria A. Dancy, "Environmental Constraints on Crude Oil Production in the United States," in *Energy Supply in the 1990s and Beyond*, Proceedings of the 11th Annual International Conference of the International Association for Energy Economics, Caracas, Venezuela, 26–28 June 1989, pp. 96–100.

13. All values are computed from data in American Petroleum Institute's *Basic Petroleum Databook*, January 1989, Section V, Table 11 and have been reflated to 1987 dollars.

14. Lifting costs are taken from the U.S. Energy Information Administration, *Performance Profiles of Major Energy Producers* (1987). Enhanced oil production statistics are given in *Oil and Gas Journal*, 18 April 1988, p. 46.

15. Offshore drilling tends to cost over six times as much per foot as onshore.

16. U.S. Department of the Interior, *Federal Offshore Statistics* (Washington, D.C.: Mineral Management Services, 1987).

17. Alexander G. Kemp and David Rose, "Comparative Petroleum Taxation," *Petroleum Economist* (February 1983): 53–55.

18. Dahl, "Worldwide Comparison." Unit costs are computed as oil wells drilled plus a portion of dry wells times cost per well times 1.66 to account for all other costs divided by reserve additions for oil. Well data are taken from American Petroleum Institute et al., *Joint Association Survey of Drilling* (Washington, D.C.: U.S. Government Printing Office, 1980, 1986, 1987), Table 1. Reserve additions are taken from U.S. Energy Information Administration, *U.S. Crude Oil, Natural Gas, and Natural Gas Liquid Reserves* (Washington, D.C.: American Petroleum Institute, 1988), p. 75.

19. Any negative costs resulting from negative reserve additions have not been included in the averages. Negative reserve additions occur when reserve adjustments have caused proven reserves to fall by more than production. There is an upward bias in costs since no allowance has been made for associated gas, while the bias from the random nature of discovering and measuring reserves in any given year is unknown.

20. U.S. Department of the Interior, "National Assessment of Undiscovered Conventional Oil and Gas Resources," (Washington, D.C.: U.S. Geological Survey and Mineral Management Services, Open File Report 88-373, 1987), p. 111.

21. Dahl, "Worldwide Comparison."

22. In-ground costs for OPEC 1987 are aggregated from individual country costs given in ibid. Country costs were computed given the following algorithm from M. A. Adelman and M. Shahi, "Oil Development-operating Cost Estimates, 1955–85," *Energy Economics* no. 11, 1 (1989): 2–10. Cost per barrel in ground (C) equals oil wells drilled plus a portion of dry holes (W) times 1.66 to account for nondrilling expenses divided by production per year per well (bw), all multiplied by the depletion rate ($d = .1$) or $C = W \times 1.66 \times d/(bw)$. For in-ground cost estimates for individual foreign countries for 1987, see Dahl, "Worldwide Comparison," and for above-ground estimates of historical costs by country, see Adelman and Shahi, "Oil Development-operating Cost Estimates."

DILLARD P. SPRIGGS

7. Impact of the Oil Price Decline on U.S. Oil Companies

THE 1986 oil price decline had severe repercussions throughout the U.S. oil industry. U.S. oil production has fallen to less than 8 mb/d. Imports are on a strong rise. Outside of the top fifteen to twenty major oil companies, the infrastructure of the industry has weakened, most notably among the independent producers and the service companies.

Almost all sectors had to contract, although the downstream business enjoyed rising margins from the sharp reduction in raw material costs. The companies that were able to endure the necessary concentration learned a lot about how to operate more efficiently. Therefore, as prices recovered in 1987, the major companies in particular were well positioned to operate in a lower price environment than most anyone had thought possible in the early 1980s. They proved that again in 1988, when prices slumped once more in the second half of the year.

This chapter reviews briefly the industry's status in the heady days of escalating prices leading up to the mid-1980s. It then looks in detail at the repercussions of the price fall on the important sectors of the industry. The discussion closes with an examination of the industry's situation in 1987 and 1988 and the kinds of operating strategies many companies adopted. Those strategies served the major oil companies very well, but a lasting impact on the national energy situation is that drilling and exploratory efforts in the conventional onshore basins of the country have been curtailed.

The Oil Industry in the Early 1980s

Well before the 1986 crash, the U.S. oil price had weakened, falling from nearly $32 in 1981 to about $24.50 at the end of 1985. That 23 percent decline had already revealed how vulnerable certain sectors of the industry could be to less remunerative prices. Their vulnerability was magnified by the great optimism which pervaded most of the industry in the early 1980s and was responsible for levels of activity which could only be sustained at a price of about $30 per barrel. The high optimism extended to the banking industry, where loans were granted with uncommon quickness on the assumption of future price escalation even though it was already apparent that prices were actually beginning to decline.

The number of wells drilled in the United States rose to a record of 91,600 at the peak of the boom years in 1981, financed in large part by the eagerness of many banks. How many of these wells were based on good geology and other sound characteristics with respect to drilling prospects will never be known, but surely many were borderline cases. An early and famous victim of the optimism was the Continental Illinois Bank, which boosted its loan portfolio from $18 billion to $31.5 billion in the three years ending in 1981, in large part by making ever-increasing advances to the oil industry. Anxious to become recognized as an important oil lender, it accepted loans originated and laid off to it by the tiny and loosely controlled Penn Square Bank of Oklahoma City. Penn Square grew so fast on the back of loans to oil prospectors that it could not, or did not, keep records of some of its loans.[1] Continental Illinois's record keeping was better, but its investigations of the purpose of borrowings and the bona fides of the debtors were casual.

The denouncement came with the collapse of Continental, the nation's eighth largest bank, which had to be rescued in 1984 with $4.5 billion from the Federal Deposit Insurance Corporation (FDIC). To lead the resurrected bank, the FDIC turned to John Swearingen, the recently retired chairman of Amoco. Swearingen thought he was prepared for the worst when he accepted the assignment, but once in the job he was surprised at how bad things were. He told friends, "These bankers didn't know what they were doing."[2]

With financing so readily available, it was likely that it would be used. In 1979 the capital and exploratory expenditures of a group of fifteen independent producing companies were equal to 97 percent of their revenues.[3] They could only spend as much as they took in by adding borrowed funds to internally generated cash. Debt came to dominate the independents' balance sheets, and servicing of it caused profits to decline sharply. However, those trends were largely ignored because of the prevailing mindset—the funds would be used to establish new reserves that at future escalated prices would pay off the debt, at which time profits would expand handsomely.

What was going on in the United States was repeated in Canada. There Dome Petroleum, the industry's largest independent company, the leading producer of natural gas liquids, and a principal oil and gas producer, expanded at a fast rate, too, tapping the resources of each of Canada's leading banks as well as New York banks and Europe's top lenders. Dome Petroleum's downfall followed the acquisition of Hudson's Bay Oil and Gas from Conoco and public shareholders. That move boosted its debt to an unsupportable level even though the price of oil was above $26 per barrel. The policy of the worldwide banking consortium turned from accommodation to anguish when the banks could not be assured of interest payments on their loans, not to mention the return of principal. In the spring of 1983, they engineered the replacement of Dome's management, which had led the company for more than twenty years, with new leadership under J. Howard Macdonald, former treasurer of Royal Dutch/Shell.

Warning Signals

Developments like those at Continental Illinois and Dome Petroleum were warning signals for everyone of possible further trouble ahead. The major oil companies were aware of these and other warnings in the years leading up to 1986. Early on Royal Dutch/Shell, as an example, based one of its planning scenarios on a $15 world oil price, but it kept the scenario to itself lest it be thought that it was predicting that sort of future.

The result was that the spending profligacy of the independent segment of the U.S. oil industry was not mirrored among the major oil companies. They maintained debt at or near its prevailing level, which in most cases was relatively low, unless they engaged in a major acquisition or had to defend themselves against takeovers, as did Phillips Petroleum and Unocal. Capital and exploratory spending for the search for and development of new production were held close to the limits of internal cash flow from operations consistent with the need to cover dividends and carry out other corporate spending. That still permitted a high level of spending on exploration and production operations in the United States and overseas (Table 7.1).

Some New Strategies

In spite of the high level of spending, most companies experienced difficulty in adding to their domestic reserves from fresh exploration. Managements became increasingly interested in other avenues than conventional exploration to achieve growth. They concentrated on two tactics: the buy-back of their own stock, which, in effect, resulted in the purchase of reserves at a moderate price, and the acquisition of reserves through property purchases or mergers with other companies.

Exxon revealed that it felt its opportunities for finding new reserves were constrained when it initiated a massive stock repurchase pro-

Table 7.1. Capital and Exploratory Spending of Fifteen Major Companies (in millions of dollars)

	1982	*1985*
Exploration and dry hole expense		
United States	6,162	6,380
Overseas	4,600	4,190
Total	10,762	10,570
Exploration and production of capital expenditures		
United States	16,871	17,616
Overseas	10,554	10,435
Total	27,425	28,051

Source: Annual reports of each company.
Note: Companies included are Amerada Hess, Amoco, Arco, British Petroleum, Chevron, Conoco, Exxon, Marathon, Mobil, Occidental, Phillips, Shell Oil, Standard Oil, Sun, and Texaco. Exploration and dry hole expenses are expensed against the income statement in the year incurred. Capital expenditures are capitalized and written off over time.

gram. At the beginning of 1982, there were 1,737 million shares of Exxon stock outstanding. By the end of 1988 that number had been reduced by 508 million shares after the expenditure of $13.6 billion, $4.2 billion in 1988. The program is continuing.[4] Exxon has not completed any mergers, but during the past four years it spent approximately $3.3 billion on the purchase of oil and gas reserves in the United States, Australia, Canada, and elsewhere. Early in 1989, the company's 70 percent owned Canadian affiliate did move to take over another company; it agreed to acquire Texaco Canada for $3.2 billion.

Shell Oil, the company with about the best record of adding to domestic reserves through conventional means at a low cost, has been quite active in buying proved properties. It spent about $2.3 billion on these purchases in the 1984 to 1987 period. Shell itself was taken over in 1985 by parent Royal Dutch/Shell, which acquired the one-third minority interest in the company held in the hands of the public. (For the principal mergers that were completed between 1981 and 1985, see Table 7.2.)

The hallmark of the period after 1981 leading up to 1986 was the striking difference in strategies between independent producing companies and the major companies. Independents went heavily into debt in order to press the search for domestic reserves. The majors, on the other hand, recognized that it was becoming increasingly difficult to find new reserves in the United States; several of them succeeded in acquiring reserves through mergers or property purchases at relatively low prices. The independents weakened themselves while the majors had gotten stronger on the eve of the 1986 price crash.

Table 7.2. Principal Mergers, 1981–1985

	Value ($ Billion)	Date	Est. Price per Barrel
DuPont/Conoco	7.5	8/81	3.71
USX/Marathon	6.3	3/82	3.55
Occidental/General Service	4.1	12/82	4.94
Phillips/General American	1.2	3/83	6.63
Texaco/Getty	9.9	2/84	4.22
Chevron/Gulf	13.2	5/84	4.45
Mobil/Superior	5.8	9/84	5.62
Royal Dutch/Shell	5.2	6/85	5.28

Source: Various issues of *International Petroleum Finance*.
Note: Price per barrel represents Goldman Sachs estimates of the consideration paid for oil equivalent barrels with natural gas converted to crude oil on the basis of 6Mcf = 1 barrel.

Fall in Cash Flow

The effect of the price crash in 1986 on cash flow was immediate. At a crude oil price of $12 per barrel, as an example, the cash flow from a barrel of crude oil falls to less than $6, as shown in Table 7.3, about half the cash flow which results from a price in the mid-1920s.

The major oil companies responded to the new situation by reducing their total capital and exploratory expenditure budgets for 1986 by 32 percent from 1985 outlays to $31.6 billion, as shown in Table 7.4. Efforts to reduce operating costs, including personnel cuts, received top priority along with the budget reduction. Management performance was scrutinized carefully. British Petroleum's early reaction was to unseat the top management at 55 percent owned Standard Oil and to place two of its top executives in control. A reorganization of operations followed. The direction of lower forty-eight exploration was shifted almost entirely to the Gulf of Mexico and spending in the lower forty-weight was slashed. Arco was also quick to streamline its lower forty-eight exploration and production operations. It cut its inventory of undeveloped acreage by more than half to 5.3 million acres at the end of 1986, and during the year it sold nearly eight hundred fields which it classified as high cost, ridding itself of 42 million barrels of liquid reserves and 335 billion cubic feet of gas reserves. Most other major companies undertook similar moves to those of Standard and Arco to rationalize their U.S. operations. Outside the United States and Canada, the international operations of the majors were not subjected to the scalpel of large cutbacks, with exploration expenses hardly touched, as noted later.

The reduction in spending by the major oil companies carried over

Table 7.3. Cash Flow from a Barrel of Crude Oil (in U.S. dollars)

Price	26.00	12.00	18.00
Operating costs	5.00	4.50	4.50
Taxes	10.50	1.70	4.50
Net cash flow	10.50	5.80	9.00

Source: Conoco.

Table 7.4. Capital and Exploratory Expenditure of Leading Oil Companies (in millions of U.S. dollars)

	1985	*1986*	*1987*	*1988*
Amerada Hess	699	217	348	730
Amoco	5,306	3,181	2,979	3,561
Arco	3,962	979	1,650	2,240
BP America	2,962	2,283	1,708	2,650
Chevron	4,035	3,018	2,800	3,300
Conoco	1,725	1,451	1,459	1,965
Exxon	10,793	7,219	7,136	7,510
Kerr-McGee	384	301	237	362
Marathon	1,163	764	781	871
Mobil	3,513	2,976	2,798	3,901
Occidental Petroleum	1,102	979	613	1,067
Pennzoil	599	233	195	227
Phillips Petroleum	1,284	801	877	911
Shell Oil	4,453	2,842	2,624	3,331
Sun	1,859	1,167	1,203	1,891
Texaco	2,670	2,197	2,234	2,613
Unocal	1,707	991	1,009	1,270
Total	46,348	31,599	30,651	38,300

Source: Annual reports of each company.

to 1987, expenditures falling another $1 billion to $30.7 billion. This was followed by an increase in outlays to some $38 billion in 1988.

Independents' Ranks Thinned

The effect of the 1986 price fall on the independent-producing sector of the oil industry was not just severe, it was disastrous for many of the participants. Developments in 1986 confronted them squarely with issues of survival, not solely with the big decline in cash flow that also affected the major oil companies. In case after case, the high debt burden of the independents could not be satisfied and the only alternative was selling out or declaring bankruptcy. Well-known companies that found selling out the best strategy included

Aminoil, American Quasar, Energy Reserves Group, Celeron, Lear Petroleum's producing function, Monsanto Oil, Petro-Lewis, Inexco, and Universal Resources. Estimates made by Diamond Shamrock indicate that of forty-five major independents with assets of less than $4 billion but more than $1 billion which existed in 1982, only thirty-four remained at the beginning of this year.

The most important privately owned independent organization, that of Herbert and Bunker Hunt, was forced into bankruptcy. However, in the end it will survive in a much slimmed down form. But that is not the fate of hundreds of other private independents. A barometer of what happened to them is that the number of paying members of the Independent Petroleum Association of America dropped to less than 1,600 in 1987 from around 6,000 in 1981.[5]

Advantages of Integration

The eclipse of many independent-producing companies was sealed by the price slump, which suddenly rendered the exploration and production business a marginal or even a money-losing activity. Integrated companies, on the other hand, had the fortune of possessing substantial downstream operations. Just as suddenly, as a result of Saudi Arabia's 1986 shift to netback pricing, these operations became big money winners, piling up record refining and marketing earnings in that year as producing profits vanished.

Although the integrated companies were worse off from a total earnings standpoint, their cash flows remained large even if they were lessened. Standing at the apex of the industry at the time of its worst period of business in fifty-five years were the leading international companies, particularly Exxon and Royal Dutch/Shell. Because of their strong downstream presence outside the United States, these companies profited handsomely from the large refining margins made possible by netback crude. And they had still another telling advantage over everyone else in the industry. They fared better than their U.S. counterparts in the upstream business because the high tax rates on foreign production absorbed much of the price decline.

Integration gave several of the major companies "the power of the purse," which enabled them even in the worst of times to purchase producing properties from others that needed to replenish their cash. As the price of oil recovered and gathered strength late in 1986 and 1987, they could even think of carrying out a merger as big as the megamergers that occurred before 1986. Thus, British Petroleum, Amoco, and Exxon's affiliate Imperial could, as described below, spend around $15 billion for acquisitions of large operating entities in 1987 and 1988.

Impact on the Service Industry

If the revenues and cash flow of the oil companies had to be reduced, it was certain that the suppliers of the oil tools and services required to conduct exploration and production would suffer. It is a given that contract drilling and oil service operations are cyclical businesses, oscillating with changes in oil and gas prices between peaks and valleys. The valleys are deepened by overbuilding during peaks, particularly in drilling rigs, where freedom of entry is easy. But the decline which these businesses suffered after 1985 is the greatest that at least two generations of management experienced.

In December 1981, 4,530 drilling rigs were at work in the United States, many drilling on prospects that, as indicated earlier, could only be justified on the most optimistic price projections, if then.[6] In that year the revenues of twenty-two publicly owned companies rose to a record peak of $37 billion. From there it was downhill for all companies, although the trend began relatively slowly. By early 1985, however, with the number of active rigs having fallen to less than two thousand and contract rates as well as service prices being cut due to overexpansion, revenues of the group of twenty-two companies declined to $22 billion and several companies operated in the red.

In 1986 rig rates collapsed (down to below $7,000 a day from as high as $40,000 in 1981) while prices of drilling bits fell around 40 percent from the previous year. No company escaped. Schlumberger, the most successful of them all, incurred no less than a $1.7 billion loss after major write-offs. The aggregate loss of the twenty-two companies was $3.5 billion, down from the peak of their prosperity of $4.2 billion profits in 1981.

However, one man's problems can be another's gain. The service companies' reduction in fees and prices lowered oil company costs of finding and developing oil. Each dollar of expenditure accomplished much more work than it had one or two years earlier. One company's experience—that of Maxus Energy Corporation shown in Table 7.5—was repeated throughout the industry.

The contract drilling service companies are still trying to work their way out of the difficulties through stringent cost-cutting measures, major write-offs, amalgamation, and consolidation. The weekly average of the number of rigs at work in the United States hit a record peak of 3,970 in 1981 and fell to 963 in 1986. In 1987 and 1988 it was

Table 7.5. Exploration Cost Comparison

	Then	Now	% Decrease
Drilling			
Anadarko Basin—two 11,000-ft wells in same field, drilled 11/85 and 11/86	19.50 $/ft	11.75 $/ft	40
Powder River Basin, Wyoming—two 10,000-ft wells in same field, drilled 5/86 and 11/86	$200,000	$109,000	46
Gulf of Mexico, main pass area—two wells to 6,000′ late 1985 and late 1986	$780,000	$487,000	38
Seismic			
Gulf of Mexico—group shoot data in 1983 vs. data shot in 1986	100 $/mi	20 $/mi	80
Alabama Norphlet play data offered 1984 vs. late 1986	1,500 $/mi	1,300 $/mi (offering price) 700 $/mi (negotiated price)	53
Land			
Powder River Basin, Wyoming—analogous leases bought 1984 vs. late 1986	75 $/acre	10 $/acre	60

Source: Maxus Energy Corporation.

no better than 937.[7] Early in 1989 the number of rigs employed fell to less than eight hundred.

Bankruptcies claimed such companies as Global Marine, a major offshore contract driller, and Smith International, a leading drilling bit manufacturer. Reorganization plans were carried out at each and one was being set in motion at Reading and Bates, an offshore driller, early in 1989. Gearhart Industries, NL Industries, and Western Company of NA were taken over by others. Two of the biggest companies in the industry, Baker International and Hughes Tool, merged.

The Industry in 1989 and a Look to the Future

The profile of the major participants in the oil industry has changed considerably in the aftermath of the oil price fall from the peak of 1981 to the trough of 1986. The major companies have survived fairly well intact, but consolidation has reduced the number of stand-alone companies. Cash flow of these companies is substantial and their financial position is strong. The independent producing sector of the industry is much reduced, particularly among the private operators, but the number of publicly owned entities has also been cut back.

Consolidation in the industry continues. The power of the purse that the majors were able to retain marked British Petroleum's expenditure of $7.8 billion in mid-1987 to acquire complete control of Standard Oil. The takeover of Standard gave BP the benefit of access to its large cash flow of about $3 billion. Previously, BP received only its share, or about 55 percent, of Standard's dividend of $660 million. About a year later in the summer of 1988, BP added to its North Sea position by spending $3.9 billion to acquire Britoil, thereby gaining 600 million barrels of liquid reserves and 3 trillion cubic feet of natural gas reserves.

At the beginning of September 1988, Amoco completed its $4.2 billion merger of Dome Petroleum in Canada, which it had initiated in early 1987. It gained 3 trillion cubic feet of natural gas reserves and 212 million barrels of crude oil and natural gas liquids. For these and other assets, Amoco's acquisition cost was slightly higher than $4.00 per oil equivalent barrel, according to the company. It also obtained 8.7 million net acres of exploratory acreage, primarily in the Beaufort Sea and Western Canada.

The sliding profitability of the oil business plagued Tenneco so much in 1986 and 1987 that it decided in 1988 to withdraw from oil and gas production. Its decision was greeted with keen interest by many other companies. By packaging its diverse properties in several separate entities and putting them up for bidding, it obtained $7.3 billion for the reserves late in 1988. The assets sold and the buyers of the different packages are shown in Table 7.6.

Future Exploration and Production

Exploratory and development activity is now being evaluated very rigorously by all the participants—independents, majors, and banks. Among the majors, the conclusion that the search for oil in most onshore areas will be unrewarding is widespread because discoveries, if and when made, are so small. Standard Oil reduced its inventory of undeveloped acreage by about half in mid-1986, charging off $410 million against income. At the time, the chief executive of the company, Robert B. Horton, said, "It is our belief, based on detailed studies, that the average explorer in the lower 48 states *can simply no longer succeed.* The margin he would recover would be negative" (italics added).[8] Mobil, one of the industry's most obvious users of advertising space in the national news media, frequently makes known

Table 7.6. Assets Sold and Buyers of Tenneco

Company	*Asset*	*Price (millions of dollars)*
Chevron	Gulf of Mexico	2,600
Amoco	Rocky Mountains	900
Mesa	Mid-Continent	715
Arco	California Heavy Oil	700
American Petrofina	Gulf Coast	600
Royal Dutch/Shell	Columbia	500
British Gas	Other Foreign	195
Conoco	Norway	115
Mobil	Louisiana Refinery	560

Source: Announcements of individual companies.

in its messages that is is placing "increasing emphasis on international exploration where tomorrow's larger and more profitable new fields are likely to be found."

This preference for international exploration has been reflected in the spending trends of the major companies since 1985. Dollars spent on exploration in the international circuit by the major oil companies have been reduced much less than U.S. exploration expenses. A recent survey by the Department of Commerce of capital expenditures points to significant increases in outlays in 1988 and 1989 over 1987 spending. The survey includes all U.S. companies and all segments of the oil business, of which spending on exploration and production operations is by far the largest (see Tables 7.7 and 7.8).

The reduced rate of spending on domestic exploration has prevented any recovery in the utilization of U.S. drilling rigs, as indicated earlier. As a result, total well completions in 1988 averaged only about half those in 1985. Oil well completions are estimated to have fallen much more sharply than natural gas wells, or by roughly 60 percent over the three years compared with about 35 percent for natural gas.[9]

Well-servicing activity reflected moderate recovery early in 1988, according to the American Petroleum Institute, but reversed direction after crude oil prices retreated again. Well servicing averaged about 30 percent lower than in 1985 over the year. And employment in oil and gas extraction dropped another 10 percent in 1988, on the basis of the institute's estimates, to about 420,000, or 30 percent below 1985.

The inevitable result of the three-year drilling setback was a marked slump in U.S. crude oil production, with little likelihood of recovery

Table 7.7. Exploration and Dry Hole Expenses of Fifteen Major Oil Companies

	1985		*1986*		*1987*	
	$ Million	%	*$ Million*	%	*$ Million*	%
United States	6,380	60.3	4,336	53.3	3,098	51.8
International	4,190	39.7	3,798	46.7	2,879	48.2
Total	10,570	100.0	8,125	100.0	5,977	100.0

Source: Annual reports of companies.

Table 7.8. U.S. Petroleum Industry Spending Abroad (in billions of dollars)

	Actual	*Estimated*	
	1987	*1988*	*1989*
Developed Countries			
Canada	2.0	2.6	2.6
Europe	4.1	5.5	5.8
Australia	0.7	0.8	0.8
Developing Countries			
Latin America	0.6	0.7	0.9
Africa	0.6	0.8	0.9
Middle East	0.5	0.4	0.4
Australia	1.1	0.3	2.1
International	0.2	0.3	0.5
Total	9.8	13.0	14.0

Source: U.S. Department of Commerce, *Survey of Current Business* 68, no. 10 (October 1988): 26–32.

any time soon. Production of crude oil averaged 8.1 mb/d in 1988, the lowest level since 1976. At year's end, production was down to 7.9 mb/d, reflecting an acceleration in the rate of decline. At that rate, production was 1.2 mb/d below what it had been in early 1986. Based on its analysis of the accelerated rate of decline, the American Petroleum Institute in January 1989 estimated the decline rate is now 400,000 b/d per year.[10]

Production of crude oil in the lower forty-eight has descended to 6.1 mb/d, the lowest level since 1950. Its fall has been partially masked by growing production in Alaska (see Table 7.9). However, output from Prudhoe Bay will commence its own decline late in 1989. Alaska is not likely to be able to offset falling lower forty-eight production with increased flow from its wells in the near future.

Table 7.9. Crude Oil Production in Alaska and Lower Forty-Eight States (in thousands of barrels per day)

Year	*Alaska*	*Lower Forty-Eight*	*Total*
1973	198	9,010	9,208
1977	464	7,781	8,245
1980	1,617	6,980	8,597
1981	1,609	6,963	8,572
1982	1,696	6,953	8,649
1983	1,714	6,974	8,688
1984	1,722	7,157	8,879
1985	1,825	7,146	8,971
1986	1,867	6,813	8,680
1987	1,962	6,387	8,349
1988	2,017	6,112	8,129

Source: Energy Information Administration, *Petroleum Supply Monthly* (January 1989): 6.
Note: Preliminary data.

The nation's oil supply includes not only crude oil but also natural gas liquids production, which has averaged about 1.6 mb/d in recent years. Between 1981 and 1985, the period of high drilling activity, combined production of crude oil and natural gas liquids increased from 10.2 mb/d to 10.6 mb/d. However, by 1988 it had fallen close to 1 mb/d to 9.8 mb/d.

Over the last three years, the largest eighteen oil companies shared in this decline to the tune of 270,000 b/d. The remaining producers in the industry saw their production fall 520,000 b/d, as shown in Table 7.10.

The availability of significant production of natural gas liquids still permits the United States to meet more than 50 percent of its petroleum requirements from its own sources, a higher ratio than that of most major oil-consuming countries. Nevertheless, oil imports rose more than 2 mb/d during 1986 to 1988 and now constitute 42 percent of domestic consumption (Table 7.11).

The prospects for reversing recent increases in imports appear limited. The major oil companies are able and willing to explore in frontier areas, the only remaining places where large geological structures exist. The most important of these potential areas are the deep waters of the Gulf of Mexico, offshore California, Alaskan waters, and the Arctic National Wildlife Refuge. Exploration in these areas requires long time periods, particularly in Alaska. Even after discovery, development would take up to five years, or even longer,

Table 7.10. Net Crude Oil and NGL Production in the United States (in thousands of barrels per day)

Company	*1981*	*1985*	*1986*	*1987*	*1988*
Amerada Hess	82	72	69	69	68
Amoco	437	401	394	394	409
Arco	540	648	659	665	674
BP America	727	783	803	854	858
Chevron	342	566	546	507	484
Exxon	752	768	761	756	760
Kerr-McGee	25	29	29	27	26
Marathon	166	160	155	142	137
Mobil	316	352	342	338	336
Occidental Petroleum	5	79	85	72	64
Pennzoil	36	35	34	32	31
Phillips Petroleum	125	154	124	113	110
Shell Oil	514	530	571	558	527
Sun	217	195	171	159	155
Tenneco	99	101	104	101	93
Texaco	381	640	590	540	524
Unocal	159	163	162	161	163
17 companies	5,075	5,796	5,725	5,589	5,526
All others	5,155	4,754	4,515	4,361	4,234
Total United States	10,230	10,590	10,240	9,950	9,760

Source: Annual reports of each company and Energy Information Administration.

particularly as new technology will be required to bring on production.

Access to offshore California and the wildlife refuge on the North Slope of Alaska is surrounded by significant political and legal problems which have been compounded by the *Valdez* oil spill in 1989. Environmental concerns over exploration in both areas are likely to remain heightened for a long time. The day when the drill bit begins to turn on prospective oil acreage in the wildlife refuge has undoubtedly been pushed back. Considering the extended period that is required to develop discoveries in and around the Beaufort Sea, it is likely to be beyond the year 2000 before production from the refuge could be established even assuming it is opened to oil explorers and discovery follows—by no means a foregone conclusion in spite of the promising geological environment.

The independent producers generally are unable to explore in frontier areas, given the high costs of most projects. Their work is likely to be concentrated in the lower forty-eight, the major basins of which now appear capable of yielding measurable reserves of natural gas but not of oil.

Table 7.11. Domestic Oil Supply and Imports (in millions of barrels per day)

	1985	*1986*	*1987*	*Estimated 1988*	*Forecast 1989*	*Forecast 1990*
Crude Oil Production	8.95	8.67	8.35	8.14	7.86	7.63
Alaska	1.83	1.87	1.96	2.02	2.02	1.96
Lower forty-eight	7.15	6.80	6.39	6.12	5.84	5.67
Natural gas liquids production	1.61	1.57	1.60	1.62	1.62	1.62
Total crude oil and natural gas liquids	10.59	10.24	9.95	9.76	9.48	9.25
Imports	4.95	6.00	6.60	7.34	7.74	8.06
Crude oil	3.08	4.08	4.60	5.04	5.54	5.81
Refined products	1.87	1.92	2.00	2.30	2.27	2.25

Source: Energy Information Administration, U.S. Department of Energy; forecast 1989 and 1990 from the base case forecast in "Short Term Energy Outlook," DOE/EIA (89/1Q) (January 1989).

Conclusion

On the eve of the nineties, the condition of the U.S. oil industry is much different from what it was when it entered the eighties. Significantly reduced are the ebullience and financial strength of the independent sector of the industry. The service industry is still sorely tested, although enough units and companies remain to accomplish whatever work there is to be done. The majors have emerged from the trials of the past three years in strong financial shape for the most part, the value of integration having been proved in the price declines of 1986 and 1988. The companies have been transformed in the sense that they are fewer in number due to mergers and acquisitions (with some substantially larger as a result) and because they are operated more efficiently than at any time in decades. As their interest in operations abroad, particularly exploration, has widened, so, too, must they accept the fact that two of their competitors—Royal Dutch/Shell and BP—have expanded their U.S. base by absorbing entirely Shell Oil and the Standard Oil Company. John D. Rockefeller might have turned over in his grave.

For the nation the evolving oil situation, particularly the well-recognized lack of any important oil discoveries onshore lower forty-eight, may give rise to renewed concerns about this country's overall energy position. It is not hard to project a further substantial decline in domestic oil production beyond what the Department of Energy is already forecasting for the next few years. Even a recovery in oil prices that leads to more exploration and drilling is not apt to affect materially future decline in oil production, although it is likely to yield significant natural gas discoveries. The prospects are for continued expansion in imports of foreign oil. The likelihood that the level and proportion of exploration dollars directed abroad will increase will reinforce this trend.

Notes

1. The saga of Penn Square Bank and the aura of the get-rich-quick syndrome prevailing in Oklahoma before the bank's astounding failure in July 1982 were brilliantly recounted in Mark Singer, *Funny Money* (New York: Alfred A. Knopf, 1985), p. 222.
2. Comment to this writer, 17 June 1985.
3. See *International Petroleum Finance,* 30 July 1980, p. 6.
4. Jack Bennett, Exxon's chief financial officer, told security analysts in early 1988 that the stock purchases up to that time "involved buying proved crude-oil-equivalent barrels for the equivalent of less than $3.40 a barrel." See Exxon report entitled *Briefing for Security Analysts,* 8 March 1988, p. 21.
5. Interview with IPAA official.
6. *Drilling Contractor* (December 1988/January 1989): 6.
7. Ibid.
8. Standard Oil Company, *Second Quarter Report to Stockholders* (1986), p. 11.
9. *Monthly Statistical Report* 12, no. 12 (December 1988), American Petroleum Institute, Washington, D.C.
10. Ibid.

WILFRID L. KOHL

8. Oil and U.S. National Security

THE oil price collapse in 1986 caused a dramatic decline in U.S. oil production. As described in the chapter by Dillard Spriggs, the independent oil companies and the oil service industry were hit by a major recession. Companies were forced to reduce their budgets for exploration and production. The U.S. oil drilling rig count decreased about 50 percent, portending a falloff in the volume of future reserves that will be found. Banks which had made energy loans came under heavy pressure, and several of them failed. It was not long before cries from the Southwest for U.S. government assistance began to be heard in Washington. Meanwhile, several government and industry studies pointed to a likely rapid rise in U.S. oil imports as a potential threat to America's energy security in the 1990s.

This chapter reviews the development of U.S. energy security policy following the oil price collapse and the congressional and political debate through the end of 1990. It then explores the role of oil imports in U.S. national security and policy options for the future. The analysis shows that oil imports are only part of the security problem, which is international in scope and rooted in the overall stability/instability of the world oil market at any given time and its vulnerability to future supply disruptions and price shocks.

The U.S. Government Response to the Oil Price Collapse

On a trip to Louisiana in September 1986, President Reagan called for a government-wide study to examine the implications of the depressed state of the U.S. oil industry and the falloff in U.S. oil production. An interagency high-level study was subsequently organized under the chairmanship of William F. Martin, who had moved a few months earlier from his post as executive secretary of the National Security Council (NSC) staff to become deputy secretary of energy under Energy Secretary John S. Herrington. No stranger to energy issues, Martin had been a key assistant to the executive director of the International Energy Agency in the late seventies before joining the Reagan administration, where he dealt with energy (especially the transatlantic dispute over the Soviet-European gas pipeline) at the State Department, and then at the NSC. Martin had already seen the need for a broad analytical study of the U.S. energy situation placed squarely in its proper international context.

After almost six months of meetings, staff work, and interagency debate, the Martin group produced a lengthy report in March 1987, *Energy Security—A Report to the President.*[1] Because of the difficulty of obtaining agreement among some fourteen participating government agencies, the final report was issued solely on the responsibility of the Department of Energy (DOE). One of the most comprehensive U.S. government energy studies ever conducted, the report analyzes all the major domestic energy sectors within the context of the international oil market. It concludes that rising dependence on oil imports from the insecure Persian Gulf will make the United States more vulnerable to market disruptions or OPEC price manipulation in the nineties. Projections show imports increasing from 5.2 mb/d (about one-third of U.S. consumption) in 1986 to between 8 and 10 mb/d by 1995 (about one-half of U.S. consumption).

While a number of policy options are mentioned in the DOE study as to how to deal with this future vulnerability, none are specifically recommended, presumably because of lack of agreement in the interagency group. However, the disadvantages of an oil import fee to the American economy and to some of our allies were stressed in an appendix. Another appendix emphasizes the costs of a gasoline tax. The report advocates the operation of free market forces. But it does lean toward tax and fiscal measures to stimulate domestic oil supply. Opportunities are pointed out for expanding U.S. natural gas and coal use (the latter with due attention to the environment as facilitated by new clean coal technology). A buildup of strategic oil stocks by the United States and its allies is encouraged to strengthen Western capabilities to deal with future oil market disruptions.

The DOE report is based on two oil price scenarios for the next decade: (1) a lower price case under which oil prices would remain at about $15 per barrel until 1990, after which the world oil price would rise gradually to $23 per barrel by 1995 (the corresponding rate of U.S. economic growth would be 2.7 percent annually); and (2) a higher price case which assumes an increase in the world oil price to $23 per barrel in 1990 and $28 per barrel in 1995 (which yields a U.S. annual economic growth rate of 2.5 percent). Meanwhile, studies by two industry groups—the American Petroleum Institute and the National Petroleum Council—and a study by the Office of Technology Assessment, a congressional agency, all issued within a few months of each other, reached essentially the same conclusion regarding increased future U.S. dependence on oil imports.[2] According to the press, Secretary of Energy John Herrington sent the *Energy Security* report to the president in mid-March 1987 with a cover letter recommending a 10 percent tax credit on exploratory drilling at oil prices under $27 per barrel and a more generous depletion allowance for domestic oil and gas producers.[3]

The DOE report and its implications were debated at a White House meeting of the Economic Policy Council (EPC), chaired by Treasury Secretary James Baker and attended by President Reagan, in late March 1987. Representatives of various government agencies disagreed as to which policy actions to pursue. Siding with DOE were the other national security agencies such as the Defense Department (which was concerned about defense of the Persian Gulf and having adequate oil supplies in an emergency), the CIA, the NSC, and part of the State Department. A counter-coalition of consumer-oriented or free market agencies included the Office of Management and Budget, which insisted that any new policy actions be revenue neutral, the departments of Commerce and Labor, and the Council of Economic Advisors. Having just fought successfully for a major tax reform, the Treasury Department opposed reopening the question of new changes in the tax law. The Interior Department was reported to favor the opening up of the Arctic National Wildlife Refuge (ANWR) to exploratory drilling and more liberal rules for leasing on the Outer Continental Shelf (OCS), but Secretary Donald Hodel was absent from the meeting.[4]

Following the EPC meeting, President Reagan decided to instruct DOE and other agencies to continue to monitor the situation in the domestic oil industry and the world oil market and to send a letter to Congress recommending action on a number of measures previously proposed and on a few relatively mild new measures.[5] The uncertainty of predictions of future oil prices, which by then had recovered to around $17 per barrel, along with the recent action on tax reform, were major factors in the president's decision to propose limited actions at that time while leaving open the possibility of further actions in the future.

The earlier proposals that were reaffirmed included repeal of the windfall profits tax, comprehensive natural gas decontrol, approval of the Department of the Interior's five-year offshore oil and gas leasing plan, exploration and development in an environmentally sound manner of the ANWR, and nuclear power plant licensing reform and other actions to help revitalize nuclear power. Of these earlier proposals, the Fuel Use Act was repealed in 1987 and the windfall profits tax in 1988 (the latter was attached to the Trade Bill, which passed in the summer of 1988).

Unwilling to reopen basic issues already considered in tax reform, the president did ask Congress to consider small technical changes: increasing the net income limitation of the percentage depletion allowance from 50 to 100 percent per property, and repealing the transfer rule to permit use of percentage depletion for proven oil properties purchased by an independent producer from a major producer.

The president also announced the administration's willingness to support a fill rate of 100,000 b/d of the Strategic Petroleum Reserve (SPR) in order to achieve the goal of 750 million barrels of oil by 1993, provided that Congress finds budget offsets to cover the costs of the higher fill rate. He indicated that he was reducing the minimum bid requirement for federal offshore leases from $150 per acre to $25 per acre to encourage exploration and development.

The American oil industry and the oil-producing states in the Southwest were not completely satisfied with the president's response. While Congress considered the items in the president's letter, some members favored more aggressive action. Senator Lloyd Bentsen of Texas sponsored a bill in the late spring of 1987 which would have amended Section 232 of the Trade Expansion Act to require the DOE to project U.S. oil import dependence three years in advance. If import dependence exceeded 50 percent, it would then have required presidential action to hold dependence at or below 50 percent.[6]

Attached as an amendment to the Trade Bill, the Bentsen bill was defeated on the Senate floor on 1 July 1987 by a vote of 55 to 41. The administration, which did not favor encumbering the president with new requirements, opposed the bill. Other congressional efforts to gain support for an import fee were even less successful, since they were viewed in partisan or regional terms. The more numerous oil-consuming states, which stand to benefit from lower oil prices, have generally been unwilling to assist the oil-producing states.

Unhappy with the administration's response, in December 1987 a number of independent oil companies, grouped together in a National Energy Security Committee and in the Texas Independent Producers and Royalty Owners Association, filed a petition with the U.S. Department of Commerce under Section 232 of the Trade Expansion Act of 1962. The petition argued that U.S. national security is being impaired by imports of crude oil and refined products and urged the Commerce Department to conduct hearings and undertake a new study of the threat, which, it was claimed, puts the United States at risk in case of future supply disruptions or in a conventional war.[7]

After a year of study the secretary of commerce forwarded a report to the president in December 1988. While noting that U.S. energy security had substantially improved since 1979, the report stated that a number of concerns remain, especially vulnerability to a major supply disruption, growing out of the recent decline of domestic oil production, rising oil imports, and increasing Western dependence on insecure Middle East sources of supply. "The investigation found that the maintenance of U.S. access to sufficient supplies of petroleum is essential to our economic security, foreign policy flexibility, and defense preparedness. Given these factors, the Secretary of Commerce found that petroleum imports threaten to impair the national security."

However, the secretary recommended that no action be taken to adjust oil imports (e.g., an oil import fee) "because such action would not be cost effective and, in the long run, would impair rather than enhance national security." Referring to his program to improve energy security sent to Congress on 6 May 1987 and reviewing the energy security actions taken by his administration, President Reagan approved the finding of the secretary of commerce in January 1989 and announced "that no action to adjust oil imports under Section 232 need be taken."[8]

The president once again urged Congress to take action on several items remaining on its energy agenda, including natural gas price decontrol, the opening up of ANWR and OCS areas to exploration and development, extension of the percentage depletion allowance for independent oil and gas producers, continuing to fill the SPR to 750 million barrels, and enactment of comprehensive nuclear power licensing reform. Noting the international dimension of energy security, the president's message asserted that the United States will continue to work with its allies in the International Energy Agency (IEA) in the building of strategic oil stocks and the achievement of a more balanced energy mix.

Meanwhile, following the issuing of the *Energy Security* report, the Department of Energy supported the administration's successful effort to obtain congressional passage of the U.S.-Canada Free Trade Agreement (FTA) in 1988. The agreement will promote mutual access to energy markets and is expected to increase the availability of Canadian natural gas imports into the United States as American natural gas demand expands, which could have positive implications for reduced oil demand and imports. The FTA will also provide a stable framework for U.S.-Canadian trade in oil and uranium.

In the last year of the Reagan administration the DOE undertook several studies on issues identified in the *Energy Security* report which in essence reevaluated the free market approach in several sectors:

1. Since transportation consumes almost two-thirds of the oil used in the American economy, it is especially important that an effort be made to expand the use of alternative fuels in this sector. But it is

doubtful that the market will do this alone. DOE has been studying alternative and flexible fuel use.[9] Meanwhile, Congress passed the Alternative Motor Fuels Act of 1988, which provides limited Corporate Average Fuel Economy (CAFE) incentives for production of alternative fuel vehicles and $18 million of government support for demonstration projects.

2. DOE has completed a study of U.S. natural gas reserves that underscores the potential for expanded natural gas production in the United States.[10]

3. At the end of the Reagan administration DOE was also studying tax regimes and other oil and gas industry investment incentives in foreign countries as background for future recommendations to strengthen tax incentives for the U.S. oil and gas industry.

In the opening months of his administration President George Bush moved slowly in defining an energy policy, focusing first on the lengthy process of nominating and achieving confirmation of key officials in DOE and other agencies. Some of the president's initial ideas can be sketched, based upon his campaign speeches, the Republican party platform, and the president's budget message to Congress in February 1989. Bush has said, "We strongly need an energy policy that makes us less dependent on foreign oil." While favoring a basic free market approach, the new president appears willing to show greater flexibility as to the role of government in energy and in the protection of the environment, building on the reassessment that began at the end of the Reagan administration. Bush's own early experience as a Texas oilman undoubtedly supports his generally favorable approach to the oil and gas industry.

The Bush administration went on record early in favor of tax incentives for oil and gas exploration and for enhanced oil recovery. It has opposed an oil import fee and, at least initially, new taxes. President Bush has generally supported the leasing of federal onshore and offshore areas for oil exploration and development (e.g., ANWR) subject to appropriate environmental standards. Some increased funding of research and development of energy efficiency, conservation, and alternative fuels is expected. And a bill to provide natural gas wellhead price decontrol, initiated in the Democratically controlled Congress in spring 1989, was signed by the president.

Following his confirmation as Secretary of Energy in the spring of 1989, Admiral James T. Watkins announced an eighteen-month study based on a series of public hearings around the country concerning the nation's energy needs. The objective was to formulate a National Energy Strategy (NES), which was finally announced by the President in February 1991 and is to be followed by new legislative proposals.

An interim report issued in April 1990 summarized public comments at many of the hearings. The final NES emphasizes increased energy production and R and D and the need to improve energy efficiency through market forces. Notably absent are measures to promote energy conservation via reduction in demand.[11]

In its first year and a half the one major action of the Bush administration in the area of energy and the environment was the proposal to amend the Clean Air Act, announced in June 1989. While subsequently modified in Congress before it was passed in 1990, the new act establishes tighter controls on acid rain and toxic vehicle emissions. The new guidelines will press the oil and automobile industries to move toward reformulated gasoline and cleaner alternative transportation fuels during the next decade.

Oil and National Security: Economic, Military, and Foreign Policy Dimensions

Will greater oil import dependence pose a future threat to the stability of the American economy in case of a future oil price shock? To America's military preparedness and leadership role in the world? Are there other variables involved in energy security? In this section we turn to an analysis of U.S. oil imports and oil consumption and the likely impact of increased oil imports on components of national security.

Oil and Economic Security. Rising U.S. oil imports played an important role in the tightening of the world oil market in the seventies, setting the stage for the oil shock of 1973–74. From a level of 3.2 mb/d in 1970, net imports of crude oil and petroleum products increased to 6 mb/d in 1973 at the time of the embargo. An important factor at that time was that OPEC controlled approximately 75 percent of world oil production. U.S. spare capacity had dwindled to zero, and no spare capacity to speak of was present in Venezuela or Canada. Only the countries of the Middle East, and especially Saudi Arabia, had the ability to substantially expand production, which gave them critical bargaining leverage.[12]

Under the impact of domestic price controls, imports continued to rise to 8 mb/d or more in 1977–78, just before the second oil shock in 1979, as shown in Table 8.1. In response to the doubling of oil prices to levels over $30 per barrel in 1979–80, which reduced oil demand and encouraged fuel switching and conservation, U.S. net imports declined by half in the early eighties. However, since the collapse of oil prices in 1985–86, net imports have been rising again

Table 8.1. U.S. Petroleum, Net Imports

Year	Net Imports (mb/d)	Imports as % of Consumption	Year	Net Imports (mb/d)	Imports as % of Consumption
1970	3.2	NA	1981	5.4	33.6
1972	4.5	27.0	1982	4.3	28.1
1973	6.0	34.8	1983	4.3	28.3
1974	5.9	35.4	1984	4.7	30.0
1975	5.8	35.8	1985	4.3	27.3
1976	7.1	40.6	1986	5.4	33.4
1977	8.6	46.5	1987	5.9	35.5
1978	8.0	42.5	1988	6.6	38.1
1979	8.0	43.1	1989	7.2	41.6
1980	6.4	37.3	1990*	7.6	45.2

Source: U.S. Department of Energy, *Monthly Energy Review*.
* Average of first two quarters.

and reached 7.2 mb/d in 1989, or 41.6 percent of oil consumption. This amount is higher than the U.S. level of oil dependence (35 percent) at the time of the October 1973 embargo. Net imports rose to 45 percent in the first half of 1990. Looking ahead, almost everyone agrees that U.S. imports will continue on an upward curve as oil demand increases with economic growth while domestic production continues to decline (see Chapter 6 by Carol Dahl). According to the American Petroleum Institute, gross oil imports exceeded 50 percent of our oil consumption on a monthly basis twice in the first half of 1990.[13]

Imports of oil or any other commodity are not in themselves undesirable. If an internationally traded good is available from abroad at a lower price than it can be produced domestically, and this is the case of oil, imports can be justified on grounds of efficiency and the nation may generally be better off. Lower oil prices are also good for the American economy. Since the United States is a large energy consumer and importer, we need to use fewer resources for traded goods.

Why then would we want to reduce imports? One reason might be the much publicized trade deficit. In 1980 oil imports represented 32 percent of merchandise imports. At the same time petrochemicals and other products with hydrocarbon inputs (thus very sensitive to oil prices) represented 35 percent of export earnings. The U.S. Treasury expressed concern about the sizable oil import share in the trade deficit.

But since that time oil has become much less important in the trade

Table 8.2. Petroleum Consumption in Organization of Economic Cooperation and Development (OECD) Countries (in thousands barrels per day)

Year (Average)	*Canada*	*France*	*Italy*	*Japan*	*United King*
1973	1,707	2.422	2,147	5,071	2,301
1974	1,740	2,260	2,090	4,960	2,138
1975	1,718	2,136	1,940	4,502	1,872
1976	1,751	2,280	1,991	4,771	1,856
1977	1,779	2,235	1,907	5,231	1,880
1978	1,823	2,169	1,948	5,142	1,850
1979	1,893	2,385	2,013	5,480	1,930
1980	1,873	2,256	1,934	4,960	1,725
1981	1,768	2,023	1,874	4,848	1,590
1982	1,576	1,927	1,779	4,549	1,584
1983	1,486	1,891	1,727	4,365	1,518
1984	1,491	1,838	1,633	4,574	1,822
1985	1,485	1,725	1,687	4,365	1,634
1986	1,506	1,772	1,697	4,391	1,637
1987	1,563	1,789	1,819	4,450	1,603
1988	1,664	1,797	1,836	4,752	1,697
1989	1,763	1,856	1,940	4,981	1,752

Source: U.S. Department of Energy, *Monthly Energy Review*.
Note: OECD (Europe) consists of Austria, Belgium, Denmark, Finland, France, Greece, Iceland, Ireland, Italy, Luxembourg, the Netherlands, Norway, Portugal, Spain, Sweden, Switzerland, Tur-

picture, even though imports have recently been rising. The reason is that the value of oil imports has been relatively stable due to the decline in oil prices in the mid-eighties while overall imports and exports have risen dramatically. By 1988 the oil import share of overall merchandise imports had declined to less than 10 percent. The main cause of the U.S. trade deficit is to be found in macroeconomic conditions, especially in the federal budget deficit, which has driven up interest rates and the value of the dollar.[14]

The economic reasons for considering a reduction in oil imports lie elsewhere in what is known as the oil import premium. The first problem with oil import dependence arises because the source of the incremental barrel of imported oil is the OPEC cartel, not a competitive producer in a free market. As can be seen from Table 8.2, the United States is a very large consumer and purchaser of oil. The United States consumes more oil than any other country or region, including Organization of Economic Cooperation and Development (OECD) Europe, and it imports more than any other country—only slightly less than OECD Europe combined. Indeed, the United States consumes just under half of total OECD oil consumption and is responsible for about 25 percent of total world oil imports of crude oil and products.[15] Thus, the volume of U.S. oil imports is likely to

nited States	West Germany	OECD (Europe)	OECD (Other)	OECD
17,308	2,915	14,521	1,006	39,612
16,653	2,612	13,708	1,056	38,117
16,322	2,515	13,059	999	36,600
17,461	2,708	13,813	1,068	38,864
18,431	2,837	13,795	1,123	40,359
18,847	3,048	13,963	1,117	40,892
18,513	3,073	14,670	1,090	41,646
17,056	2,707	13,634	1,072	38,595
16,058	2,449	12,515	1,080	36,269
15,296	2,323	12,069	1,000	34,489
15,231	2,287	11,772	940	33,794
15,726	2,296	11,781	994	34,565
15,726	2,352	11,566	956	34,098
16,281	2,498	12,013	948	35,139
16,665	2,424	12,169	931	35,759
17,283	2,422	12,375	944	37,018
17,325	2,278	12,561	976	37,607

y, the United Kingdom, and West Germany. OECD (Other) consists of Australia, New Zealand,
d the U.S. Territories.

have a major impact on the world market price. As demand for cartel output is increased, the world price is generally driven up by the OPEC cartel. The price paid by all U.S. oil consumers (and consumers elsewhere in the world) will increase, not just for the importer of the marginal barrel of oil.

Economists call this hidden cost of increased import demand the "demand component" of the "import premium."[16] This phenomenon is particularly likely to occur in a tight oil market, when the cartel would tend to increase prices in response to increases in demand.[17] Given the important role of the United States as an oil importer, further growth in U.S. imports could be a major stimulus leading to higher world prices. The United States might therefore be better off importing less oil in order to maintain lower prices of the oil we continue importing. This argument is also known as the optimal tariff argument and, if the conditions hold, would be the basis for an oil import fee. One problem with the argument is that it is hard to predict whether OPEC would retaliate against a tariff or import fee by restricting further its oil production. As discussed further on, an oil import fee would also carry with it undesirable domestic macroeconomic side effects.

Another reason to be concerned about oil imports is the security argument associated with vulnerability to an oil price shock. Oil import dependence does not necessarily equal vulnerability in the security sense. Vulnerability is a measure of the ease or difficulty of adjustment—in both economic and political terms—of a country to a change in the availability or price of an important commodity. Vulnerability can exist even without dependence. Since oil is a fungible commodity in a world market, a supply disruption somewhere in that market from a country from which the United States imports no oil would still lead to a rise in the world price of oil if other countries were dependent on imports from that country and if little spare capacity existed in the world oil system. A significant disruption could lead rapidly to a price shock that would affect the entire world oil market, including the United States.

A threat to economic security that could lead to considerable economic costs exists if (1) there is some risk of disruption of supplies of a vital commodity such as oil and the possible magnitude of the disruption is sufficiently great, and (2) there is considerable dependence of the country on oil imports. If the cost of adjustment in the economy is substantial, that is, if flexibility to switch to other fuels (or other nonfuel inputs) does not exist, then a country's economic security is vulnerable.[18] The costs might come from reduced supply availability, although that aspect could be mitigated quickly in a fungible world market. Most important would be the impact of an oil price escalation. In that case the macroeconomic loss to the economy of a major disruption would have two components. The first would be the loss from a sharp oil price spike, which would require real wages to fall and capital and labor to be substituted for energy in the economy. This cost has been termed the "disruption premium" or "security premium" of imported oil. The second would be the increased wealth transfer to foreign suppliers equal to the amount spent on oil imports.

According to Henry S. Rowen and John P. Weyant, "Both effects are important: the former, economic dislocation, is responsible for about two-thirds of estimated GNP losses and the latter, the transfer to foreign producers, for about one-third of the estimated losses."[19] They have estimated that the first oil shock of 1973 reduced the U.S. GNP by about 5 percent and that the second oil shock of 1979 reduced GNP by nearly 3 percent.[20] A broader study conducted by the Stanford-based Energy Modeling Forum, which analyzed the results of a number of economic models, concluded the first oil shock reduced the 1975 GNP level by 3 to 5.5 percent and the second shock yielded a GNP loss for the United States in 1981 in the range of 2 to 4 percent.[21] An OECD study reported the real income losses to all

OECD countries due to the second oil shock at about 5 percent in 1980 and nearly 8 percent in 1981.[22]

Vulnerability associated with oil imports derives from the fact that the concentration of future supply in a politically unstable region, the Middle East, brings with it the risk of a future supply disruption and an associated oil price shock. The fact that the United States has done much in recent years to diversify its sources of oil supply and now receives the major part of its imports from Canada, Mexico, and Venezuela misses the point. The Western industrial world as a whole is still very dependent on oil from the Middle East. In 1987 western Europe and Japan received 43 and 66 percent, respectively, of their oil imports from the Middle East.[23] (U.S. imports from Saudi Arabia grew substantially since 1985 to 1 mb/d in 1988 and increased to around 1.2 mb/d in 1989.)[24]

The world oil market at the end of the 1980s continued to show excess supply and production capacity. OPEC's share of world oil production is running around 40 percent, which does not give the cartel the leverage it needs to exert control over the market. Saudi Arabia has given up its role as swing producer, and the economic and budgetary needs of many cartel members frequently lead them to cheat on their production quotas. Non-OPEC production continues to grow slightly, though it may fall off in the 1990s. But what of the future?

Western imports from the Gulf region are likely to grow in the nineties as oil demand increases and non-OPEC production stabilizes and eventually declines, for the Persian Gulf is the locus of two-thirds of the world's oil reserves and enjoys the lowest cost of production. While a truce has replaced war between Iran and Iraq, political rivalry between these two traditional enemies will not cease. The civil war in Lebanon, the longstanding Arab-Israel dispute, the threat of a spread of Islamic fundamentalism, the possibility of terrorism or internal revolution in any number of countries, and the expansionist ambitions of Iraq—any or all of these factors will continue to make the Persian Gulf a very volatile region.

U.S. vulnerability to supply or price shocks could be reduced if flexibility existed in the economy on both the supply and the demand sides to switch to other fuels. The problem is that there are no available easy substitutes for oil in the transportation sector, which currently absorbs about 63 percent of U.S. oil consumption. While DOE, Congress, and some of the automobile companies have begun to focus on this problem and the need to develop flexible fuel vehicles, progress in this area will not be rapid given the lack of infrastructure and market incentives.

As long as major parts of the industrial world are heavily dependent on Persian Gulf oil, and as long as oil enjoys a prominent place in the U.S. energy economy (42 percent of total energy demand in 1989), U.S. vulnerability to the price effects of an oil market disruption will be real. The integrated nature of the world oil market means that energy security for the United States is integrally linked to energy security for our allies.[25] Indeed, it actually makes more sense to think about vulnerability of the world oil market to disruptions, not just one nation. A rising level of U.S. oil imports, coupled with modest oil demand growth in other OECD countries and more rapid growth in oil demand in less-developed countries, will over time increase the vulnerability of the global oil market.[26] That vulnerability is, however, reduced somewhat by the U.S. Strategic Petroleum Reserve and the oil emergency stocks of other countries, especially if they were to be used in a coordinated effort such as the United States has encouraged within the International Energy Agency.

Military Oil Requirements

In peacetime the U.S. military uses between 2 and 3 percent of national consumption of crude oil and petroleum products. Of this requirement, totaling just under 500,000 b/d, about two-thirds is composed of jet fuels for military aircraft. In case of military conflict this requirement would increase by a factor of 2 or 3 or even more in the event of a longer conventional war, which would necessitate priority support for defense-related industries. Still, the military requirement represents a relatively small slice of total domestic oil consumption.[27] About two-thirds of the current military demand comes from the continental United States or Canada, and the remainder is purchased abroad. Most of the incremental military demand in wartime would likely be overseas in one or more war zones. The Department of Defense does maintain large petroleum stocks in the United States and abroad.

In case of a large-scale defense emergency, the military has access to a range of nonmarket mechanisms for procuring oil and other fuels for defense needs. Critical would be the invoking of the Defense Production Act and the Energy and Security Act, which would allow the government to divert petroleum supplies from civilian needs to the military and associated priority defense industries. In a worst-case mobilization, defense needs might require 4 to 5 mb/d, which certainly could be met from current U.S. domestic production of around 10 mb/d.

Nevertheless, it has been pointed out that the DOD is vulnerable to oil market disruptions. Since normally it purchases oil competitively on the open market, there may be political reluctance to invoke the Defense Production Act, for example, in a smaller-scale conflict or in the opening phase of a conflict. There may also be problems of effective administration and coordination with the oil industry. "As a result of the oil embargo of 1973, for example, the DOD lost about 10 percent of its daily petroleum requirement, with the heaviest impact in jet fuels. The Defense Production Act was eventually invoked to provide supplies, but not without significant curtailment of naval and air force operations. DOD petroleum stocks were not drawn down." Defense operations could also be reduced by an oil price escalation associated with an import curtailment, which might rapidly stretch Pentagon budgets for military operating costs.[28]

Nevertheless, on balance it is difficult to become overly concerned about military oil requirements, which represent such a small part of total national oil demand. In the words of one knowledgeable DOD official, "So long as the military requirement can be supplied at all, it can be supplied as a priority, meaning the risk of import vulnerability falls on the non-defense economy."[29]

Implications for Foreign Policy and Geostrategic Interests

While difficult to quantify, there may be important indirect international political costs of increased oil imports to the pursuit of U.S. foreign policy and strategic goals. First, rising oil imports, especially from politically volatile or unreliable countries, can lead to a loss of flexibility in foreign policy and thereby to a weakening of America's leadership role. For example, in April 1986 American bombers attacked Tripoli to punish Libya for its support of international terrorism. As former Defense and Energy Secretary James Schlesinger has pointed out, that action might not have been possible had the United States been importing oil from Libya at the time (which the United States was not) under conditions of a tighter oil market instead of a general world oversupply of oil that existed.[30] Similarly, the United States might not have been able to ban imports of Iranian oil and to encourage other nations to join in that ban if U.S. oil import dependence had been higher and world oil supplies tighter than they were.

Second, since many of our allies in western Europe and Japan are even more dependent on oil imports than the United States, increased U.S. import dependence could lead to competition with our allies in a tighter world market for higher-priced oil and resulting political strains in alliance relations. Perhaps the starkest example occurred in the fall of 1973 during the first oil shock when several western European countries acted unilaterally and collectively within the European community to establish closer ties to Arab oil-producing countries at a time when Secretary of State Henry Kissinger was attempting to negotiate a Middle East peace settlement. After the International Energy Agency was established at U.S. initiative in 1974 to coordinate Western energy policies and share oil in an emergency, strains continued in the middle and late seventies when European leaders repeatedly pointed out that the United States was not doing enough to reduce its oil imports, which approached 50 percent in 1978. Rising U.S. oil imports in this period exerted pressure on the world market and helped set the stage for the oil price run-up in 1979. Political pressures exerted on President Carter by Chancellor Helmut Schmidt and other Western leaders were instrumental in Carter's 1979 decision to decontrol U.S. oil prices.

Third, the presence of two-thirds of the world oil reserves in the Persian Gulf has given that region enhanced strategic importance, as is well known. While U.S. dependence on Gulf imports declined in the eighties, it is likely to increase in the nineties as demand grows and non-OPEC production stagnates or falls. U.S. and allied dependence on oil imports from the Gulf became a focal point for U.S. foreign and security policy in 1980 when the U.S. Rapid Deployment force was created to deter or resist Soviet intervention in the area. Later, with the escalation of the Iran-Iraq war and attacks by both sides on oil tanker traffic, and in response to a request from Kuwait, the United States and several European nations increased their naval deployments in the Gulf to protect the shipping lanes. This involved considerable economic cost and posed serious strategic risks, as evidenced in the attack on the USS *Stark* in 1987, with attendant loss of life, and the erroneous action by the USS *Vincennes* in the summer of 1988 in shooting down an Iranian civil airliner. The most dramatic manifestation of U.S. strategic interest in the Gulf is the deployment there of over 400,000 U.S. and allied land, air, and naval forces in the fall of 1990 following Iraq's invasion and occupation of Kuwait, with the announced objective of defending Saudi Arabia and forcing an Iraqi withdrawal. An implicit objective of the U.S.-led U.N. coalition is to prevent Iraq's domination of Gulf oil supplies and future oil pricing given the critical role of oil in the world economy.

In the larger arena of East-West competition, energy and oil imports also play a role. Of the two superpowers, the Soviet Union, the world's largest oil and natural gas producer, is presently self-sufficient in energy. The United States is a major oil importer, as we have noted. The Soviet Union exports oil and gas to the West, and these exports are a vital source of Soviet hard currency earnings. During the Cold War it was argued, "U.S. strategic interests are adversely affected by rising Soviet hard currency earnings from oil exports. Such earnings enhance Soviet economic performance and military posture. Soviet expansionism and activism in the third world, which represent very costly commitments, also could be bolstered by higher hard currency earnings."[31] However, in the post–1989 context of East-West cooperation, U.S. assistance is being considered to help increase Soviet oil and gas production as a step toward diminishing world dependence on OPEC oil.

Conclusions and Policy Options

As the foregoing analysis has attempted to demonstrate, U.S. vulnerability in economic terms to oil supply disruptions and price shocks does not depend on U.S. oil imports alone, although they are part of the story. Vulnerability depends collectively on worldwide oil supply and demand, on total U.S. oil demand (consumption), on U.S. imports, on fuel (or other input) flexibility in the U.S. economy, and on the presence of and political will to use emergency oil stockpiles.[32]

First, worldwide oil supply and demand determine oil prices. If world demand increases to a level where it can absorb most of the available oil production capacity, OPEC benefits from a position of sufficient leverage on the world oil market to behave like an effective cartel. In this case it can gradually raise prices or, in case of a disruption in supply, effectively hold onto higher price levels set off by the spot market in the disruption, as occurred in 1979. The United States is faced with higher oil prices whether or not it imports oil.

Total domestic U.S. consumption is very important, both as a component of world demand (and hence an important variable affecting the world price) and because it determines the impact of an increase in oil prices translated as higher input costs in the national economy. This produces a shift in the size and composition of national output. There are also distributional effects as wealth is transferred from consumers to producers.

U.S. oil imports are of more limited importance, chiefly as a determinant of the income or wealth transfer effect. If we buy more imported oil, we send more money abroad, which reduces the income available to create domestic demand. It also contributes to the trade deficit, although as we have shown oil now plays a much reduced part in the U.S. trade picture.

The economic repercussions of a supply or price disruption could be mitigated substantially if the United States had flexibility to shift from oil to other fuels in a crisis. Such flexibility exists to a very limited degree today in the industrial and utility sectors but practically not at all in transportation, which is the largest oil-consuming sector.

The U.S. Strategic Petroleum Reserve contains more than 580 million barrels of oil, sufficient if oil imports were totally cut off to replace them for almost three months. West Germany and Japan also possess significant oil stocks. A start has been made in the International Energy Agency to develop a system of coordinated release of emergency oil stocks by those countries which have them. This system of insurance reduces vulnerability of the oil market to some degree.

In designing policies to reduce U.S. energy vulnerability it is important to maintain a global perspective appropriate to a world oil market subject to world supply and demand. Considerations of U.S. military oil requirements and foreign policy and strategic flexibility strengthen the case for not permitting U.S. oil imports to climb to intolerably high levels. However, the economic costs of restraining further growth in oil imports would be considerable. It is difficult to define precisely what the magic level is. It is not realistic to try to place a cap on U.S. oil imports at, say, 50 percent of oil consumption, a level which will inevitably be attained and surpassed in the next few years. Even at that level of imports, West European and Japanese dependence on oil from the Persian Gulf will undoubtedly be much greater than U.S. dependence. The United States should maintain as much diversity as possible in sources of imports. But the level of total U.S. oil consumption itself is important. Equally important is the amount of consumption and imports of our allies, along with oil demand and imports in the Third World, in determining the availability and the world price of oil—hence the vulnerability of the world oil market to disruptions. Third World oil demand should be watched especially, since it is tending to grow at a more rapid rate.

Turning to the supply side, the prospects are that non-OPEC countries, including the United States, will find few if any giant oil fields in the future but will discover smaller, higher-cost oil reserves. There is some disagreement concerning the precise outlook for the United States. The recent DOE *Energy Security* report and industry studies all conclude that U.S. oil production will decline and that oil imports will rise significantly. Under this scenario U.S. production might decline from the current level of about 9 mb/d to around 6 mb/d by the end of the century. However, well-known petroleum geologist William Fisher of the University of Texas argues that since the United States still has considerable oil reserves, U.S. production levels could remain stable for some twenty-five years at prices below $25 per barrel if discovery and production of oil could be made more efficient with improved technology.[33] The United States should pursue this approach by encouraging further advances in exploration and production technologies and enhanced oil recovery. But it does seem clear that the United States, which is a mature oil province, is left today with small, scattered reserves that will require extensive drilling and other relatively expensive techniques to produce. The United States will remain a high-cost producer. Thus, the oil price will be critically important in projecting future U.S. supply. This leads us to an important paradox. U.S. oil supply security will be enhanced by higher oil prices, but the overall health of the U.S. economy will be more robust under lower oil prices.

As pointed out earlier, the Reagan administration response to the oil price decline was minimal. A consensus did not yet exist on what government should do about the problem. However, a broad range of policy options is being discussed, and the Bush administration has begun to propose some actions. We conclude with an overview of policy actions that are being debated, or should be considered, to deal not only with the likelihood of declining U.S. oil production and rising imports but with the other international factors that could increase the vulnerability of the world oil market. The purpose here is not to analyze in depth or to endorse particular options, but to point out the range of proposals and offer a preliminary assessment.

Supply Side Measures to Increase U.S. and Other Non-OPEC Oil Production

Tax Incentives. Following the 1986 oil price decline and a further dip in oil prices in 1988, the Bush administration has proposed in its fiscal 1990 budget message a number of tax incentives to stimulate domestic oil and gas production. Specifically, four kinds of incentives were sought: (1) a 10 percent credit on the first $10 million in intangible drilling cost (IDC) expenditures per year per company and a 5 percent credit thereafter; (2) a 10 percent capital credit for enhanced oil recovery methods; (3) elimination of 80 percent of intangible drilling costs as a preference item; and (4) modification of existing depletion rules that discourage the transfer of marginal wells to independent producers.[34]

Tax incentives do cost money to the government treasury but do not carry with them the damaging side effects to the domestic economy or our trading partners of an oil import fee.[35] Changes in both federal and state tax codes could provide greater incentives for petroleum exploration and production without impairing trade competitiveness. More rapid asset depreciation schedules could also be considered. In general, this policy approach seems a sensible one to encourage domestic production and thereby reduce the need for imports. The option buys time until further changes can be effected in the nation's energy mix.

The repeal (1988) of the windfall profits tax removes a burdensome accounting and reporting requirement from oil firms and a disincentive for future investment under higher prices. The fall 1990 budget compromise included some but not all of the Bush proposals outlined above to provide tax incentives for independent producers: a new 15 percent tax credit for tertiary oil recovery projects; reduced alternative minimum tax for intangibles and marginal properties; higher net income limitation for percentage depletion; higher percentage depletion for stripper oil; and repeal of the property transfer rule. Not included was a tax credit for exploratory intangible drilling costs. These energy security incentives were estimated to be worth $2.5 billion over the period 1991–95.[36]

Encouraging Research and Development. Lower oil prices have led to cutbacks in industry R and D expenditures, especially in the oil service sector. Increased government assistance for R and D, along with government cosponsorship of industry/university cooperative projects, could, over time, contribute to enhanced oil production. A key area is petroleum geosciences, where more support is needed; for example, in enhanced oil recovery, seismic analyses, improved understanding of the potential for adding new reserves in older fields, and offshore production technology.[37] Any R and D advances that can be made would presumably also be available to improve production technologies and techniques in other non-OPEC countries. The cost of increased government expenditures could be made up by revenues from a gasoline tax (discussed below).

Improving Leasing Policies and Access to Federal Lands. Two major prospective areas for oil exploration are the federal waters off California on the OCS and the ANWR in Alaska. For several years a congressional moratorium prohibited leasing for oil and gas development of the OCS off California, and the resistance of state and environmental groups continues to delay leasing there. The *Valdez* oil spill in the spring of 1989 makes the issue even more sensitive. President Bush decided in February 1989 to delay two California OCS lease sales. In June 1990 he broadened this decision by prohibiting new drilling off the coasts of California, Florida, and other states.

The Department of the Interior believes there is a 46 percent chance of recovering 3.5 billion barrels of oil from ANWR. President Bush has spoken out in favor of development of ANWR while protecting the environment. But action by Congress will be necessary to permit exploratory drilling on the ANWR coastal plain, where there is concern about potential danger to the porcupine caribou herd and other species.[38]

The disaster of the *Valdez* oil spill clearly postponed any congressional action on ANWR or the OCS until after 1989 as attention focused on how to strengthen environmental regulations related to oil production and transportation. Assuming that improvements can be made in environmental assessment and crisis intervention, it would seem prudent to proceed with exploratory drilling in ANWR, which may represent the last giant U.S. oil reserve. If ANWR is found to contain recoverable reserves and can be developed with due attention to preserving the environment, it could make a moderate contribution to reducing U.S. oil imports in the years ahead.

There are also other technical improvements that could be made in the Department of the Interior's leasing program such as reducing minimum bonus bids and rentals, extending lease terms, reducing royalties, and exploring alternative ways of awarding leases, which could make oil exploration more attractive.

An Oil Price Floor. If oil prices drop again and remain at a low level, say below $14 per barrel, an oil price floor could be set by the government to provide a more stable climate for future U.S. oil industry investment. Such insurance would be valuable given the high costs of U.S. production and in order to avoid the disruption caused in the industry by the price collapse of 1986. In effect, an oil price floor would be the same as a variable import fee that brings the price up to a target level and then phases out if the world price exceeds the target price.[39] Ideally, such a policy should be adopted also by our allies in the International Energy Agency for maximum effectiveness, as the IEA did in 1977 when it set a minimum safeguard price of $7 per barrel.

Demand Side Measures to Constrain U.S. and World Oil Demand

A Gasoline Tax. Most industrial countries have much higher taxes on gasoline than does the United States. For years U.S. gasoline taxes averaged $.29 per gallon, whereas European gasoline taxes range in the neighborhood of $1.50 to over $3.00 per gallon. A higher U.S. gasoline tax would decrease gasoline demand and overall U.S.

oil consumption and would reduce oil imports. Both factors would exert downward pressure on world oil prices. The tax would also raise much-needed revenues for the federal treasury, some of which could support government R and D assistance to the oil industry. The disadvantages are that a tax might also reduce oil industry revenues and profits as gasoline demand declined. Nevertheless, politically such a tax may be more neutral than assistance to the oil industry. The federal gas tax was increased by 5 cents per gallon in 1990.

Conservation and Energy Efficiency. Following the oil shocks in the seventies a number of government incentives were put in place to promote energy conservation. These included the 55 mph speed limit, tax incentives for insulation in buildings, the CAFE standards for new automobiles, etc. While higher energy prices were probably the main driving force for the considerable progress achieved in conservation and energy efficiency by the early eighties, the fact remains that the decline in oil prices after 1985 has led to a reduced level of commitment to conservation. Some of the government standards have expired or are no longer being implemented. Renewed standards and incentives are another policy option to consider to place some constraints on future growth in energy and especially oil demand.[40]

Expanding the Use of Natural Gas. As pointed out in the DOE report *Energy Security*, natural gas can be substituted easily for oil in many applications, thereby lessening U.S. demand for imported oil. The recent repeal of the Fuel Use Act removes restrictions on utility use of gas and allows gas to compete with other fuels in the utility market. The completion of natural gas decontrol with the 1989 amendments to the Natural Gas Policy Act of 1978, although perhaps not seen as urgent at a time when oil and gas prices were low, will provide a more stable framework for expanded gas use over the long term. So will efforts begun by the Federal Energy Regulatory Commission to open up natural gas pipeline transportation, encouraging more competition and allowing buyers to shop for least cost gas supplies. In an effort to promote expanded use of gas in this country, the U.S. Department of Energy released a study in 1988 providing a more favorable assessment of U.S. natural gas reserves.[41] The U.S.–Canada Free Trade Agreement will provide a more stable environment for access to Canadian gas supplies in the future as U.S. gas demand grows, reducing oil demand and therefore contributing to energy security.

Revitalizing Nuclear Power. The U.S. nuclear industry has been at a standstill with no new plant orders for a decade. Construction costs have spiralled as the time required to complete new plants, fed by federal licensing delays, has elongated to twelve to fourteen years (a record in the industrial world). An uncertain outlook for electricity demand growth has made it even more unlikely that utility managers, under constant pressure from public utility commissions, would order expensive new nuclear plants. Public opposition to nuclear power continues because of uncertainties relating to plant safety and nuclear waste. The industry has also been plagued by bad management.

Actions by both government and industry could help restore the nuclear option in this country, which would also assist in reducing future oil demand. Recent growing concern about the greenhouse effect and global climate change, rooted in large part in the burning of fossil fuels, could provide the context for a more favorable second look at nuclear power. However, the fundamental problems highlighted above will have to be addressed. The federal government could help by creating a more fair and efficient structure in the Nuclear Regulatory Commission (NRC), headed by a single administrator, and by passing a law establishing one-step licensing and other measures of nuclear licensing reform. It could also allocate more federal R and D funds to match industry support for the development of smaller, simpler, and safer nuclear reactor designs. As suggested in a recent study, the government could also pass legislation to allow plants to be taken away from utilities that have not been able to operate them safely.[42]

Encouraging the Use of Alternative Vehicular Fuels. The U.S. transportation sector currently accounts for more than 62 percent of all U.S. oil use. If alternative vehicular fuels could be developed and introduced into the marketplace, they could make an important contribution to reducing energy vulnerability as well as to the improvement of air quality in major American cities. The Department of Energy has been studying the use of alternative fuels in the transportation sector and the prospects for a flexible fuel vehicle, as noted earlier. Recently, the Motor Fuels Act of 1988 was passed by Congress and signed into law, providing for CAFE incentives for the development of vehicles fueled by methanol, ethanol, and compressed natural gas and $18 million of government support for demonstration projects. And the Department of Transportation announced a $35 million program for the procurement and retrofitting of buses, minivans, and vans which use alternative fuels. The Clean Air Act of 1990 sets tougher standards for mobile source emissions in the nation's smoggiest cities over the next ten years. Along with provisions for fleet vehicles and a California pilot program, the law will encourage development of alternative fuels. But much more remains to be done.

Limiting Oil Use in the Developing Countries. While projections of growth in oil demand are relatively small for OECD countries, a much greater increase in oil demand is expected in the Third World, which is likely to experience faster economic growth.[43] If the less-developed countries become overly dependent on oil, this could place considerable future pressure on the world oil market. But the situation is not totally outside of human control. With needed capital and technology, these countries could invest in nonoil energy sources and production—in gas and renewable energy sources, for example. (Coal should be less encouraged because of CO_2 emissions and the greenhouse effect.) Moreover, they need assistance to employ energy in the most efficient way possible. A multipronged strategy seems required, including loans and credits from the World Bank and other multilateral lending institutions and incentives for private investment.[44] Support for these endeavors should be an important part of U.S. energy policy.

Options to Improve Oil Crisis Management and the Energy Component in U.S. Foreign Policy

The Strategic Petroleum Reserve. The SPR currently contains more than 580 million barrels of oil, equivalent to less than three months of oil imports. The continued filling of the SPR at a somewhat faster rate (100,000 b/d) was favored by the Reagan administration and by Congress. While the Reagan administration resisted detailed planning in advance for the use of the SPR, it did announce its readiness to use part of the reserve early in a market disruption to combat oil price escalation. However, the Bush administration is more cautious, having refused to release SPR oil early in the Persian Gulf crisis in 1990 (except for a test run). The SPR provides a useful instrument to offset limited, supply and price disruptions. As oil imports increase, a larger SPR will be needed to maintain the same level of protection against an import cutoff.

International Coordination of Emergency Oil Stocks. The effectiveness of the SPR could be reduced if other consuming countries do not take steps to reduce their demand or increase supply in a crisis. Since 1984 the United States has been working in the International Energy Agency to promote the development of government-controlled emergency stocks in other industrial countries and to coordinate use of such stocks in a crisis. Presently Germany and Japan are the other countries with sizable stocks. These efforts should be continued and support for this most important activity of the International Energy Agency reaffirmed. However, while coordinated stock drawdown in a crisis could be essential to the effective use of the SPR, given the integrated nature of the world oil market, we should not oppose *demand restraint measures* favored by other IEA partners in a crisis *if they are effective*. We should seek a coordinated IEA strategy of emergency stock drawdown coupled with effective demand restraint.[45]

Oil and Other Aspects of U.S. Foreign Policy. The Middle East will become even more critical to U.S. and world oil supplies in the future. The pursuit of diplomatic policies that promote stability in the area and economic interdependence with the United States will be important in reducing the likelihood that the Gulf countries could again decide to use an oil cutoff as a political weapon. It also requires U.S. leadership to control and reduce conflicts in the region, a task that will be especially challenging in the wake of Iraq's invasion of Kuwait.

In order to reduce U.S. dependence on Middle East oil and reduce OPEC's leverage in the world oil market, it is in U.S. interest that oil supplies remain diversified as much as possible among non-OPEC countries in the world. This will lessen the impact of a disruption in any single source of supply and reduce the ability of OPEC to use oil as a political weapon. An option to be considered here is what might be done to increase oil exploration, production, and trade in the Western hemisphere, including the use of large reserves of heavy oil in Venezuela. In the new East-West context, technical assistance to help the Soviet Union expand its oil and gas production should also be carefully examined.

The oil shock of 1990 and Persian Gulf war have raised energy security once again to the top of the nation's agenda and highlighted the dangers of increasing Western dependence on Persian Gulf oil. It is up to the Bush administration and Congress to review energy security issues and decide on further actions to slow future growth in U.S. oil consumption and oil imports. Such actions should include conservation and further diversification of U.S. energy use. U.S. policies should also include attention to the equally important international dimensions of the energy vulnerability problem; namely, how to expand non-OPEC oil production, diminish worldwide oil demand, especially in the developing countries, and promote international buildup and coordinated use of emergency oil stocks and/or serious demand restraint measures in a future energy crisis.

As usual, much will depend on what happens to oil prices. The challenge for the future will be for government to strike a balance between the free market and further actions to ensure American and international energy security in a world of lower oil prices. If the past is any guide, we need to be prepared to face energy uncertainty in the future as U.S. and other non-OPEC production declines and the world oil market tightens. Environmental concerns will reinforce the need to reassess U.S. energy policy, but that is another story.

Notes

The author wishes to thank without naming them a number of friends and colleagues in government and industry for background discussions of issues and events discussed here during the last several years. He is grateful to Dr. Michael A. Toman of Resources for the Future for constructive comments on an earlier draft.

1. U.S. Department of Energy, *Energy Security: A Report to the President* (March 1987), 240 pp. plus appendices.

2. See American Petroleum Institute, *Domestic Petroleum Production and National Security* (December 1986); National Petroleum Council, *Factors Affecting U.S. Oil and Gas Outlook* (February 1987); and Office of Technology Assessment, *U.S. Oil Production: The Effect of Low Oil Prices* (July 1987).

3. See *Washington Post*, 18 March 1987; *Wall Street Journal*, 19 March 1987.

4. The above information is based on interviews with Reagan administration officials.

5. See President Reagan's letter to Congress, 6 May 1987.

6. The Bentsen bill was titled the Energy Security Act of 1987, SS 1440.

7. Petition under Section 232 of the Trade Expansion Act of 1962, as Amended, for Adjustment of Imports of Crude Oil and Petroleum Products, submitted by Enserch Corporation, on behalf of the National Energy Security Committee and by the Texas Independent Producers and Royalty Owners Association, 1 December 1987.

8. Quotations are from the statement by President Ronald Reagan released by the White House on 3 January 1989.

9. See, for example, U.S. Department of Energy, *Assessment of Costs and Benefits of Flexible and Alternative Fuel Use in the U.S. Transportation Sector*, Progress Report One: *Context and Analytical Framework* (January 1988); U.S. Department of Energy, *Assessment of Costs and Benefits of Flexible and Alternative Fuel Use in the U.S. Transportation Sector*, Progress Report Two: *The International Experience* (August 1988).

10. U.S. Department of Energy, *An Assessment of the Natural Gas Resource Base of the United States* (May 1988).

11. U.S. Department of Energy, *Interim Report, National Energy Strategy—A Compilation of Public Comments*, DOE/S-0066P (1990); U.S. Department of Energy, *National Energy Strategy: Powerful Ideas for America* (1991).

12. See Steven A. Schneider, *The Oil Price Revolution* (Baltimore: Johns Hopkins University Press, 1983).

13. See *The Energy Daily*, 20 June 1990.

14. For the foregoing analysis I am indebted to a presentation at the Washington Energy Conference at SAIS, 15 March 1989, by W. David Montgomery of the Congressional Budget Office, and to data supplied by David Curry of the U.S. Treasury.

15. BP Statistical Review of World Energy (June 1988).

16. The following discussion is based on Harry G. Broadman and William W. Hogan, "Oil Tariff Policy in an Uncertain Market," *Discussion Paper Series*, John F. Kennedy School of Government, Harvard University (November 1986), E-86-11; and American Petroleum Institute, "Domestic Petroleum Production and National Security," 30 December 1986, sec. 2. For a more complete economic framework, see Douglas R. Bohi and W. David Montgomery, *Oil Prices, Energy Security and Import Policy* (Baltimore: Johns Hopkins Press for Resources for the Future, 1982).

17. The demand component of the import premium can be illustrated by a simple numerical example. "Suppose U.S. oil imports stand at 4 mmbd and the world price is $18 per barrel. Now assume that if U.S. oil import demand was to increase by 1 mmbd, the price would rise to $20 per barrel. The total oil import bill for the United States would increase from $72 million to $100 million per day. The total cost to the United States of each of the additional one million barrels would be $28 per barrel. However, the private cost of the marginal barrel would be only $20, the market price. Thus, in effect, the United States pays a premium of $8 for the last barrel imported" Broadman and Hogan, "Oil Tariff Policy in an Uncertain Market."

18. Hanns W. Maull in *Energy, Minerals and Western Security* (Baltimore: Johns Hopkins Press for the International Institute of Strategic Studies, 1984), chap. 1, points out that eventually economic costs could imply sociopolitical strains as a result of unemployment, a drastic reduction in production, inflation, etc., which might undermine the political legitimacy of the government or political system.

19. See Henry S. Rowen and John P. Weyant, "The Oil Price Collapse and Growing American Vulnerability," paper presented at the Energy Modeling Forum, Stanford University, September 1986, pp. 9–10.

20. Ibid., p. 4.

21. B. G. Hickman, H. G. Huntington, and J. L. Sweeney, eds., *Macroeconomics Impacts of Energy Shocks* (North-Holland: Elsevier Science Publishers B.V., 1987), pp. 9–11.

22. Sylvia Ostry, John Llewellyn, and Lee Samuelson, "The Cost of OPEC II," *OECD Observer* 115 (March 1982): 37–39.

23. BP Statistical Review of World Energy (June 1988).

24. U.S. Department of Energy, *Monthly Energy Review* (January 1989).

25. *Energy Security*, pp. 8, 25–32.

26. See Chapter 10 by John H. Lichtblau on "The Future of the World Oil Market."

27. *Energy Security*, pp. 8–9, 218–20.

28. See American Petroleum Institute, *Domestic Petroleum Production and National Security* (1986), chap. 5. For a stronger argument that a weakened U.S. oil industry following the 1986 oil price collapse now endangers national security by jeopardizing the supply of oil that would be needed in wartime, see the "Petition before the Department of Commerce under Section 232 of the Trade Expansion Act of 1962, as Amended, for Adjustment of Imports of Crude Oil and Petroleum Products," submitted by Enserch Corporation on behalf of the National Energy Security Committee and by the Texas Independent Producers and Royalty Owners Association, 1 December 1987. For a related analysis that argues for inclusion of the costs of military outlays for defense of the Persian Gulf in the import premium, see Milton R. Copulos, "The Hidden Cost of Imported Oil," unpublished paper (Fall 1988) distributed by the National Defense Council Foundation, Alexandria, Va. It should be noted, however, that these military outlays also protect the Persian Gulf oil supplies of our West European and Japanese allies.

29. From a presentation by Jeffrey A. Jones, director, Energy Policy, U.S. Department of Defense, at an International Energy Seminar at the Johns Hopkins Foreign Policy Institute, 26 January 1989.

30. Based on remarks of the Honorable James R. Schlesinger at a conference at the Brookings Institution, 15 October 1987. See the subsequent publication *Oil and America's Security*, Ed. Edward R. Fried and Nanette M. Blandin (Washington, D.C.: Brookings Dialogues on Public Policy, 1988), pp. 12–13.

31. *Domestic Petroleum Production and National Security*, pp. iv–10. For more detailed treatment, see Ed A. Hewett, *Energy, Economics and Foreign Policy in the Soviet Union* (Washington, D.C.: Brookings Institution, 1984).

32. For some of the ideas in the following formulation, I am indebted to Robert W. Fri, "New Directions for Oil Policy," *Environment* 29, no. 5 (June 1987): 16–20, 38–42.

33. William L. Fisher, "Can the U.S. Oil and Gas Resource Base Support Sustained Production?" *Science*, 26 June 1987, pp. 1631–36.

34. See *Energy Daily*, 10 February 1989.

35. The most persuasive advocate of the import fee has been Professor William W. Hogan of Harvard University, who, with other colleagues, has based his argument on the fact that oil, unlike other commodities, plays a vital role in the national economy, and the price and production of oil are controlled not by the free market but by the OPEC cartel. Purchasing imported oil involves an "import premium," discussed above, because the price paid for it by U.S. consumers does not reflect the higher social cost to the nation of dependence on insecure supplies. By reducing imports of oil, an import fee would also maintain downward pressure on the world price of oil. See Broadman and Hogan, "Oil Tariff Policy," and W. W. Hogan and B. Mossavar-Rahmani, *Energy Security Revisited*, International Energy Studies No. 2, Harvard Energy and Environmental Policy Center (1987). As pointed out in the *DOE Energy Security* study, which opposes an oil import fee, one

problem with the fee is that the higher energy prices induced by it would be costly to consumers, contribute to inflation, and depress the economy. Because U.S. manufacturers would have to pay higher energy costs, it could hurt export sales and injure U.S. competitiveness. It would cost more jobs outside the energy sector than it would create in that sector. Moreover, demands for exemptions from important U.S. allies such as Canada, Mexico, and the United Kingdom could significantly undermine the effectiveness of the fee. And the imposition of the fee could invite retaliatory protectionist measures from other countries.

36. See the summary provided in "Capital Comments," *Washington Analysis Corporation/County Nat West*, 9 November 1990.

37. Office of Technology Assessment, U.S. Congress, *U.S. Oil Production: The Effect of Low Oil Prices* (1987), summary, pp. 23–24.

38. For the analysis of the Department of the Interior recommending exploration of ANWR, see *Arctic National Wildlife Refuge, Alaska, Coastal Plain Resource Assessment*, Report and Recommendation to the Congress of the United States and Final Legislative Environmental Impact Statement, Department of the Interior (1987). Estimates of economically recoverable oil were increased in a Department of Interior *Fact Sheet* (6 February 1991). See also Office of Technology Assessment, *Oil Production in the Arctic National Wildlife Refuge: The Technology and the Alaskan Oil Context* (1989).

39. For discussion in favor of a variable oil import fee, see S. Fred Singer, "Oil Policy in a Changing Market," *Annual Review of Energy* (1987): 454–60.

40. For supporting analysis, see W. U. Chandler, H. S. Geller, and M. R. Ledbetter, *Energy Efficiency: A New Agenda* (Washington, D.C.: American Council for an Energy-Efficient Economy, 1988).

41. U.S. Department of Energy, *An Assessment of the Natural Gas Resource Base of the United States* (May 1988).

42. See John F. Ahearne, "Will Nuclear Power Recover in a Greenhouse?," Discussion Paper ENR89-06 (Washington, D.C.: Resources for the Future, 1989). For further treatment of problems of the U.S. nuclear industry, see U.S. Office of Technology Assessment, *Nuclear Power in an Age of Uncertainty* (1984).

43. Jayant Sathaye, Andre Ghirardi, and Lee Schipper, "Energy Demand in Developing Countries: A Sectoral Analysis of Recent Trends," *Annual Review of Energy* 12 (1987); 253–81.

44. See, for example, John E. Gray, W. Kenneth Davis, and Joseph W. Harned, eds., *Energy Supply and Use in Developing Countries: A Fresh Look at Western (OECD) Interests and U.S. Policy Options* (Lanham, Md.: University Press of America, 1988).

45. See Douglas R. Bohi and Michael A. Toman, "International Cooperation for Energy Security," *Annual Review of Energy* 2 (1986): 187–207.

EDWARD L. MORSE AND JULIA NANAY

9. The Oil Price Collapse: The Response of the Oil Exporters

THE oil price break of 1985–86 created serious problems for the national oil companies of the petroleum-exporting countries. These state-owned companies had to begin to function in a highly competitive commercial environment but they had almost no understanding of how competitive markets work and what is required to succeed in them. The price collapse also accelerated the massive restructuring of the international petroleum industry and the role of the national oil companies in world industry.

Among the new problems the national oil companies were confronted with in 1985–86 were how to sell enough oil to maintain their market share; how to set prices at a level which would assure that all they committed to the market would be sold; how to maximize earnings through a combination of pricing policies and the establishment of special ties with refiners; and, finally, how to use hedging tactics on the futures market to achieve a minimum price.

The major exporting countries both inside and outside OPEC developed market responses for the first time and thereby revolutionized the world oil market. New mechanisms for the pricing of crude oil were devised. Market integration re-emerged between producing and consuming countries. Futures markets became a much-used vehicle to reduce the risk of price fluctuations.

Oil market fundamentals, which underlay the competition for markets leading to the price collapse of 1985–86, are not likely to change until well into the next decade. Until the fundamental balance between global demand and production capacity improves, it is not likely that any recovery will be permanent. Any rebound in oil prices that occurs in the meantime as a result of the agreement of major exporters to share markets is likely to be tenuous.

Competition among the exporting countries is driven by their desire to maximize the net revenues they derive from what for most of them is their best source of foreign exchange. By withholding some of their production from the market, the exporters try to maintain prices at a higher level than they would receive if they maximized production. However, this usually means that one or more exporters do not maximize the value of their oil. Saudi Arabia and some of the other so-called low-population, high-reserve countries pursue the dual objectives of preserving a market alliance among the exporting countries and also keeping oil prices in the short to medium term at a level which stretches out the life of oil as a major fuel.

Exporting countries are also in stiff competition with each other. In a market burdened by overcapacity, one nation's gain in market share is another's loss. Objectives other than increasing market share are sometimes pursued, such as the desire to reduce the oil income of some of their competitors for geopolitical reasons. When a group of Middle East exporters in 1986 sought to deprive Iran of revenues needed to continue its war against Iraq or to punish Libya on foreign policy grounds, oil strategy was a useful instrument. However, as long as the overhang in oil production capacity persists, it is unlikely that nonoil objectives will be pursued for any length of time.

Soft market conditions have shifted price risk to the producer and away from the refiners and marketers of petroleum products. In this environment oil exporters need to find a way to deal with price risk. Exporters also have to find their own market niche since they lose some of their profit when they rely on traders and spot sales. Some market-based pricing principles facilitate tying refiners to long-term contracts.

While netback pricing has this advantage it also has the disadvantage of shifting all price risk to the producer. Other forms of market-based pricing allow the producer and refiner to share more equitably in the price risk. At another extreme, a few exporters have been able to establish even tighter relationships with refiners through processing deals, establishing joint ventures, and buying into downstream operations in consuming countries' marketplaces. To the degree that an exporting country is able to forge one of these special relationships, it is also better able to capture some of the downstream rents it would otherwise lose and to substantially reduce its price risks. Mechanisms have also been developed as insurance against price risk. These include hedging on the futures markets and less formally structured arrangements, making use of petroleum as a financial instrument.

Developing country exporters, however, find it difficult to take advantage of these opportunities. They need to develop staffs with sophisticated commercial skills, currently in short supply in most developing countries. In order to profit from these mechanisms, a degree of autonomy is also required in the commercial departments of national oil companies. This, however, would contradict the multiple checks and balances that are normally in place in order for governments to guard against the misuse of funds.

In the oil market of the late 1980s one exporter's gain was another's loss or lost opportunity. Exporters can increase their export revenues only by a combination of aggressive marketing and skillful pricing. An active stance must be taken in the marketplace, severely taxing the human resource capacity and skills of most oil-exporting countries.

Before 1985–86, most oil traded internationally was priced administratively by governments. This was as true of the principal non-OPEC countries (the United Kingdom, Norway, and Mexico) as it was of OPEC. The few marginal exporters who priced their exports essentially on a spot market basis rode the tails of the price setters. Administered pricing reflected the resource nationalism of the 1970s and was supported by market conditions: The balance between supply and demand was relatively tight. OPEC prices were geared to the so-called marker crude, Saudi Arabian Light (34 degree API gravity) crude oil, the most widely traded crude. Other OPEC official selling prices were set with reference to the marker, and price differences among them were based on differences in physical properties and distances from the major markets.

This system was relatively easy to maintain in the 1970s because the overall balance between supply and demand was tight and because the marker crude was priced to sell at a market-clearing price in the most distant clearing market (i.e., CIF U.S. Gulf Coast). European refiners thus were able to keep the economic rent which came from being closer to Gulf production and because they were able to use efficient very large crude carriers (VLCCs). Occasionally, auctioned oil created spot prices which tested the market clearing levels at which the world's key crude oils were being sold. High auction prices, as in 1973–74 and 1978 to 1980, facilitated the rapid upward adjustment of official prices. Soft prices, as in 1976, provided market signals to the price setters not to increase prices at all or only moderately.

The system became increasingly dependent on one exporting country, Saudi Arabia, playing the role of swing producer and was only viable so long as the Saudis were willing to continue in this role. At times, this system wreaked havoc on long-term export sales contracts. Whenever an exporting country decided to test its pricing system, it could do so only by breaking contracts and auctioning crude to the highest bidder. Administered pricing unraveled at an increasing rate from 1981 to 1985 when it was replaced by netback pricing, as a seller's market was transformed into a buyer's market.

After OPEC's meeting of December 1986, administered pricing was reintroduced on a limited basis. OPEC set a targeted average price of $18 per barrel, which was supported by a collective quota for OPEC production. In addition, a selected number of OPEC crudes was included within a basket of crudes around the price target of $18. These included Arabian Light, Dubai Fateh, Algerian Sahara, Nigerian Bonny Light, Indonesian Minas, and Venezuelan Tia Juana Light. The basket also included one non-OPEC crude whose price was wholly market-based—Mexican Isthmus. Other crudes had their selling prices established via a differential from the basket.

Administered pricing was extremely difficult to carry out in the buyer's market of the late 1980s. OPEC only accounted for about 43 percent of 1988 world crude oil production.[1] Even if OPEC members tried to administer prices, virtually all other non-OPEC crude oil sales contracts were (and are likely to continue to be) based on some clear market-based pricing principle. This means, in practice, that the only way OPEC members could assure their targeted prices was by dividing the remaining market among themselves. One exporter (Saudi Arabia) or a group of exporters (the Gulf countries) had to be willing to act as swing producers.

Another difficulty exporters had in 1987 was that buyers were reluctant to sign fixed-priced contracts that did not provide flexibility with respect to liftings. Buyers are apt to renege on contracts if the administered price for a specific crude oil and the market price for the same or equivalent crudes diverge significantly. Another drawback was that the establishment of fixed differentials under administered pricing directly affected the relative demand for one crude versus another and thus changed the value of crudes vis-à-vis one another, undermining the very effort to administer prices. Another problem was that even if the nominal price of crude oil exports were fixed, crude oil is priced in U.S. dollars, and the dollar is itself subject to fluctuations in the market. Hence, the effort by OPEC to raise prices by close to 15 percent in the winter of 1986–87 was accompanied by a fall in the relative value of the dollar. Finally, only significant crude oil exporters, whose crudes are highly valued by the market, can set a fixed price. For a small exporter, there can be a severe risk of lost market share due to the fact that buyers of marginal crude oils will not accept the price risks of fixed priced sales. Administered pricing is not viable until the market is more closely balanced and was abandoned in mid-1987. It was replaced by market-based pricing.

Auctions and Spot-Related Sales

The opposite of administered pricing is selling crude oil on a spot sale basis to the highest bidder. Auctioning has some advantages: Auctions will virtually guarantee that the crude oil will be sold and lifted, provided a sale is announced in a timely manner; auctions are useful for some exporters when markets are tight and when refiners are willing to pay a premium for an extra, or marginal, crude supply, a condition not likely to appear for many years; auctions are sometimes useful in today's market for very small producers, like Benin and a few other West African countries, which may wish to sell only one or two cargoes a month.

However, in the relatively soft market of the late 1980s auctions are only occasionally an attractive alternative for selling oil. Relying on auctions to market a country's oil would undoubtedly require holding frequent sales. This would overburden a marketing staff, which could be forced to sell on a cargo-to-cargo basis and require verification of the financial status of each lifter. Auctions would also reduce earnings in soft markets. They are frequently associated with "distress sales" of cargoes whose lifting is required in order to free storage capacity. Finally, the per barrel price of auctioned crude sales is more likely to fluctuate over the course of a year than that of alternative methods.

Crude Oil Price Linkages

Crude-linked pricing arrangements have many advantages and have grown rapidly since 1986. Some exporters now link the price of their crude oil to that of widely traded and quoted crudes, which are comparable in quality. The growth in the transparency of the crude oil markets during the past half decade has made this method feasible.

Once an appropriate linkage is found, the system is fairly easy to administer in a uniform manner, although frequent adjustments are required. It is also appropriate for a small producing country which lacks a sophisticated marketing staff. This system is market-responsive and it can facilitate the development of long-term offtake arrangements with refiners. If an appropriate linkage can be found, a refiner will be indifferent to purchasing one crude versus another, thus facilitating the refiner's long-term commitment to purchase what would otherwise be a marginal crude.

Two Latin American countries—Ecuador, an OPEC member, and Colombia, a new oil producer—have adopted this method. They link their crude oil price to that of Alaska North Slope (ANS) crude sold on the Gulf Coast of the United States, and Ecuador also uses this formula in its sales to the Far East. Ecuador's crude (Oriente) and Colombia's (Caño Limon) are similar to each other and to ANS. Their FOB prices primarily reflect differences in transportation costs. Short-term price fluctuations are minimized by using a five-day average of quoted U.S. Gulf Coast prices for ANS, the mid-point day of which is the day the crude is lifted. Other crudes traded internationally are often tied to Brent (United Kingdom), Dubai, Oman, and, in the past, Arabian Light.

Price linkages have some disadvantages, the most important of which is exemplified in Ecuador. The tying of Oriente to ANS is an ideal basis for pricing crude oil into the market via traders rather than via direct sale to refiners. The bulk of Ecuador's crude sales to the United States is ultimately sales to trading companies, which resell the crude, presumably at a profit, to refiners. This means that rents are being lost by Ecuador's national oil company, CEPE. To obtain these rents CEPE would need to seek out relatively small refiners for whom Oriente can be a significant long-term source of crude oil. A pricing mechanism would also need to be devised to capture the trading profits now being lost.

The linkage of Oriente to ANS exemplifies another disadvantage of this type of pricing mechanism, which can be found in the formulas used by Kuwait, Saudi Arabia, and other producers who link their crudes to ANS. ANS is an unusual crude oil in a special market. ANS is produced at an enormous distance from the Gulf Coast market. It is available on the Gulf Coast only because there is a glut of crude oil on the West Coast of the United States and the government forbids its export. Therefore, ANS must be sold in the East, incurring enormous shipping costs from Alaska and through the Panama Canal. It has, in short, a depressed value to begin with. The price of ANS is itself tied to two other crude oils in the U.S. market, West Texas Intermediate (WTI) and West Texas Sour (WTS). One of these crudes, WTI, is the crude oil traded on the New York Mercantile Exchange (NYMEX) for future delivery (i.e., paper barrel trading). The price of WTI, and hence of ANS and Oriente and of crudes linked to it, is open to speculative market pressures, both on the downside and the upside. Since NYMEX contracts are cleared monthly, the price of ANS is also open to special pressures at least one day a month.

Market-Basket Pricing

Tying the price of crude oil exports to a basket of crude oils or a combination of crude oils and petroleum products is a more complex alternative to a simple price linkage. This method reflects overall market conditions and smoothes out some of the volatility that may be introduced in a simple linkage.

Mexico's sales contracts, developed over the course of 1986, are a good example of market-basket pricing. Although the exact formula has changed over time, Mexico has introduced daily price settings calculated from a weighted basket of widely traded crudes. For Western Hemisphere deliveries principally to the United States, prices of both Isthmus and Maya, Mexico's two main crudes, are tied to a weighted basket of WTI, WTS, ANS, and fuel oil. For Mexico's European deliveries, the weighted basket price also includes Brent and at one time included Flotta, Urals, and European fuel oil quotations. The fuel oil component was used to adjust for the heavier or lighter characteristics of Mexico's two crudes. This procedure is relatively transparent as it is based on a set of widely traded and quoted crudes. It was, however, difficult to negotiate and involved prolonged discussions between Mexico's commercial staff and its customers. Mexico thus developed an active marketing posture and expanded its market monitoring and sales operations in the United States and Europe.

Mexico had some additional objectives in deciding to adopt market-basket pricing. It was determined to adopt a pricing strategy that would enable it to withstand the risk of losing market share, as it had in the early part of 1986. Mexico also wanted to avoid the disadvantages of netback contracts, claiming that they were destabilizing and artificial. The Mexican commercial authorities found that basket pricing was responsive to changes in the market as well as competitive, transparent, regionally uniform, and simple to manage.

Market-basket pricing has advantages over both simple price linkages and netback contracts. During times of high crude prices and low product prices, market-basket pricing provides greater value to exporters than netback contracts (which place all price risk in the hands of exporters). On the other hand, netbacks are likely to be superior to market baskets when crude oil prices are low relative to product prices.

Market-basket pricing is an attractive marketing system for a country like Mexico, which has relatively large volumes of crude oil for sale in big markets. This system has a certain amount of flexibility. If a producing country seeks to increase its market share, it can alter the basket formula. During a period of tightening supplies the producer can then reverse the procedure to extract the maximum price possible. The timing of the pricing of the crude basket can also be adjusted. During a period of falling prices, the exporter can benefit by keeping lags to a minimum, while in a tight market longer lags can result in higher revenues.

However, a system which requires flexibility to maximize prices of crude oil also requires an astute and active commercial department, not generally available in developing countries. A disadvantage of basket pricing is the lack of actively traded spot market crudes from which to calculate prices. Liquidity has often been a problem, especially in the Brent market. With an active spot market for Brent crude, stress is often placed on crude baskets linked to it. Another problem with basket pricing is that as product demands change over time, so too will the relative values of various grades of crude. However, any formula can be adjusted to compensate for such changes.

Netback Pricing

Netback contracts enjoyed a revival in 1986 when they were adopted by Saudi Arabia, and other exporting countries followed the Saudi example. Netback contracts set the price of crude oil according to the price refiners receive for their products. After processing and transportation costs as well as a set profit margin for the refiner are deducted, the FOB price of the crude oil at the export terminal is netted back.

Netback pricing has one clear advantage. Since it guarantees a profit for a refiner, it also provides a means for an exporting country to maintain or increase its market share. But it does this because it places all price risk on the exporter and absolves the buyer from any price risk whatsoever. In 1986 Saudi Arabia found that netbacks provided an extremely efficient means of regaining lost market share. By guaranteeing profitability on the processing of their crude, the Saudis were able to increase their sales very rapidly by 2 million barrels per day at the expense of other OPEC members and some of the major non-OPEC producers (principally Mexico and Egypt).

When crude oil prices are high relative to product prices, however, netback arrangements will be less attractive to oil exporters. On the other hand, when crude oil prices are low relative to product prices, the crude exporter is likely to benefit from netbacks. Netback contracts, especially the long-haul contracts from the Middle East that were negotiated in 1986, were priced on a delivered basis, frequently sixty to ninety days after lifting. This exposed the oil exporters not only to price risks associated with fluctuations in transportation costs but to the risk of product price changes during transit. These risks had previously been borne by buyers under fixed contract arrangements.

Detailed negotiations are frequently required for each netback agreement, resulting in fixing the mix of products on which the netback terms are based. In some cases, refiners find that product demand shifts and they have to convert their refineries to change the mix of light and heavy products. This can be advantageous to either the buyer or the seller, depending on the terms of the agreement.

Netback agreements have been criticized for destabilizing the oil market. If refiners are assured a profit, they have an incentive to run at full capacity, even to the point of dumping extra output on the spot market, pushing product prices down even further. Spot crude oil prices are also forced lower since refiners are not willing to pay as much for oil that does not guarantee a profit. It used to be that falling product prices due to oversupply or declining demand would reduce refiners' profit margins, leading them to cut output. The reduced output in turn would push product prices back up. This self-correcting mechanism is inoperable under netback arrangements, and no stabilizing factor appears to exist at the crude oil supply level.

Finally, netback contracts have also been criticized for being refinery-specific, which can result in separate pricing arrangements for each contract, even though the refiners involved in the contracts might be in one marketing area. One variant of netback contracts is "deemed" netback pricing, whereby the pricing terms are negotiated for a "typical" refinery in a given market area. This arrangement is simpler to negotiate and throws some additional risks (and also some potential gains) on refiners, who have a new incentive to keep their costs down.

In conclusion, the virtue of netback contracts—their ability to assure exporters market share—is significant. They provide a means of reintegrating producers with refiners and marketers. They are a noncapitalized form of integration. The main disadvantage of netbacks is in the producers' loss not only over market control but also over their participation in rents generated downstream from the refinery.

Retroactive Pricing

The retroactive setting of prices after crude oil was delivered to market became widely used for Middle East crude destined for the Far East market after the 1986 price collapse. This method was based on the buying habits of Japanese oil purchasers—their reluctance to assume the price risk on long-haul crudes and their preference for long-term contracts, for which they often are willing to pay a premium. It also reflected the special relationships that existed between some governments in the exporting countries and the buyers of their crudes. Retroactive pricing was pioneered by Oman and has also been widely used by Abu Dhabi, Dubai, Brunei, China, and Mexico, especially for sales to the Far East.

Retroactive pricing combines a formula approach (Oman and Dubai crudes have an active spot market) and a certain amount of analysis. Market signals are provided by widely traded crudes combined with information from local producers, buyers, outside consultants, and price-reporting services. Retroactive pricing is most appropriate for some Middle East suppliers to the Asian market, given the long haul involved. The exporter can gain the advantages of retroactive pricing by its decision on when to time the pricing of its crude. An average of market prices taken closer to the date of delivery than the date of loading can accomplish this. But not many crude sellers can work with the partly subjective elements required for retrospective pricing to work. It thus has limited appeal.

Processing/Tolling Arrangements

Another pricing option for exporting countries is to negotiate tolling arrangements with refiners having spare refining capacity. In a tolling arrangement, the producing country negotiates an appropriate fee with the refiner for processing its crude and storing refined product until it is sold. Given the amount of surplus refining capacity available, such arrangements have been carried out in every refining center—Singapore, Rotterdam, and the Caribbean.

A favorable arrangement for the producer would require the negotiation of a fee, on a per barrel basis, that is close to the refiner's marginal operating costs rather than to his or her average operating costs, thus guaranteeing the refiner a fair profit but preventing the refiner from overcharging the producer. A processing or tolling contract is a major step toward market reintegration. The producer country, after having arranged to have its crude oil refined, then must engage in direct negotiations with distribution companies. The producer can thereby obtain some of the downstream profits by marketing his or her own products in industrialized and, to a lesser degree, developing countries.

A number of producers, especially from the more mature oil-producing countries, have concluded attractive tolling contracts with refiners. However, tolling contracts can work only if the producer country has the human resources to engage in direct product marketing. Skilled staff is needed to negotiate a fair pricing standard for product sales. In every major refinery center there is a high level of competition in the products markets, and market standards are available for pricing products at the wholesale level. Therefore, transparent standards exist, making such negotiations much simpler than those for crude oil export contracts.

Tolling contracts are a good substitute for more complete forms of market integration. They are useful when capital resources are lacking for downstream investments abroad, when political impediments exist to undertaking such downstream investments outside one's country, or when a government seeks to capture some of the downstream rents but is reluctant to take the risks associated with direct foreign investment.

Market Reintegration

The most mature and sophisticated national oil companies have been able to go well beyond the adoption of market-based pricing principles. They have embarked on an entirely new strategy—that of buying refineries and product distribution facilities in the markets of the major consuming countries. This aspect of change since 1986—the so-called reintegration of the market—has had a major effect on the entire structure of the international petroleum industry.

Market reintegration was one of the most dramatic developments in the international petroleum industry in the latter part of the 1980s. Changes in crude oil markets in 1986 caused the major oil exporters to reassess their investment strategies as well as their approach to marketing. Their current strategies, which emphasize a reintegration of the industry, are reversing the severance of the links between the upstream and downstream segments of the industry which occurred in the 1970s.

The new pressures for reintegration of the industry stem principally from the recognition by suppliers that market conditions have undergone a radical change since 1985. Since a supply overhang exists, there is stiff competition in the marketing of crude oil and petroleum products. In the present situation of oversupply, the profitability of investments in productive capacity is a direct function of access to markets to sell the available crude. Conversely, profitability in the downstream, or refining side of the industry, depends on the high utilization of refineries and, therefore, on access to supply. Some producers now seek refineries to provide secure outlets for their oil, and some independent refiners are seeking both direct market access to distribute products and direct access to crude supplies.

Venezuela and Kuwait began aggressive downstream acquisition efforts at a time when oil prices were high and they had considerable excess cash. National oil companies now have less capital available for acquisitions, but downstream assets continue to be available and can be acquired at attractive prices. Venezuela and Kuwait have benefited considerably from foreign downstream integration, since

Figure 9.1. Selected OPEC members: variation from quota.

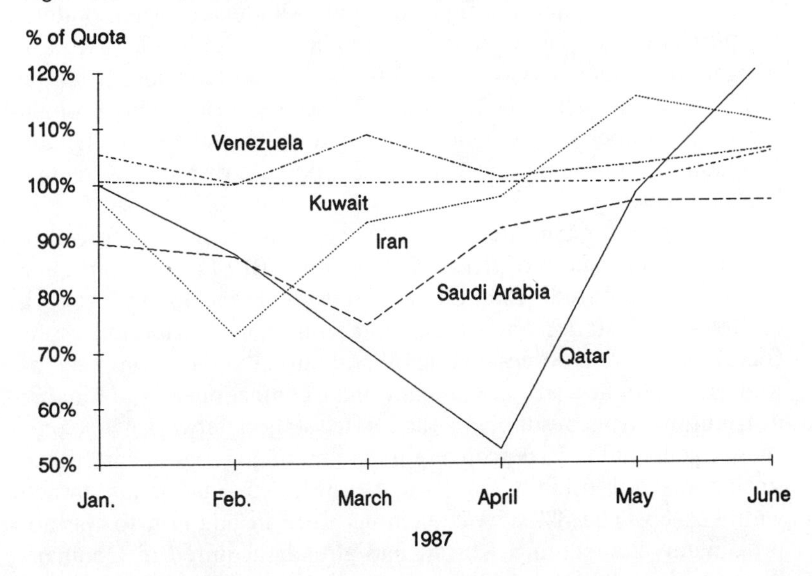

their access to downstream outlets has permitted them to hold production levels steady. While Qatar, Saudi Arabia, and Iran have been plagued by wide and erratic production swings, Kuwait and Venezuela have managed to keep oil output at or above their OPEC quotas after 1986.

Kuwait

Kuwait began integrating overseas in 1983 by buying into the European market. Paying about $120 million in cash and $175 million in crude oil, Kuwait purchased Gulf Oil's Benelux (Belgium, Netherlands, Luxembourg) and Scandinavian holdings. It acquired two refineries with combined capacity of approximately 135,000 b/d (60,000 b/d in Denmark and 75,000 b/d in the Netherlands) and more than 15 terminals and 1,575 service stations. In 1984, Kuwait purchased some additional Gulf Oil assets in Italy, giving it approximately 1,500 service stations in that country and a 75 percent interest in an 80,000 b/d refinery in Sarni.[2]

Kuwait moved on to the United Kingdom market in 1986, where it paid close to $280 million to acquire 821 gasoline stations from Hayes Petroleum Services.[3] Kuwait expanded its position in the U.K. market in 1987 with its acquisition of Ultramar's U.K. marketing operations, the Golden Eagle Petroleum Company. Comprising a commercial fuels and domestic heating oil sales network as well as 35 service stations and the right to supply 430 dealer-owned outlets, the purchase price came close to $77 million.[4] Also in 1987, Kuwait bought 53 gasoline service stations in the United Kingdom from the Soviet oil trading firm Nafta at an undisclosed price. This brought the total number of retail outlets owned by Kuwait in the United Kingdom to approximately 1,339 or 2.2 percent of the U.K. petroleum market.[5]

Kuwait's European integration strategy was further reinforced in 1987 by its acquisition of British Petroleum's (BP) Danish marketing assets for a reported $160 million. These assets included a network of 389 service stations and other marketing and distribution assets (i.e., a nationwide home-heating oil and burner service company, as well as a motor accessories company and a confectionery and tobacco distribution business supplying the service station network). The purchase also included 50 percent equity in four major coastal and fifteen smaller inland distribution depots that British Petroleum jointly owned with Texaco. The 389 service stations were in addition to the approximately 260 stations Kuwait had already acquired in Denmark from Gulf Oil in 1983. The acquisition increased Kuwait's share of the Danish petroleum market from 8 percent to 23 percent and makes it the third largest gasoline retailer in Scandinavia, after Royal Dutch Shell and Statoil. It also brought the number of Kuwait's European outlets, spread across six countries, to about 4,800.[6]

By early 1988, Kuwait had acquired a 21.6 percent stake in British Petroleum, whose shares it had begun to purchase following the British government's sale of its holdings in October 1987. Alarmed by the prospect of growing Kuwaiti control over BP, the British government directed its Monopolies and Mergers Commission to review the matter. In October 1988, the commission handed down its decision, ordering Kuwait to reduce its stake to 9.9 percent. While Kuwait was given three years to carry out the sale, BP repurchased the 11.7 percent of the shares in contention in early 1989.[7]

Kuwait is contemplating downstream operations in France, West Germany, and Spain. Spain is of particular interest but since there is considerable hostility to a sizable foreign presence in the Spanish oil industry, Kuwait has moved cautiously. It has entered Spain's downstream industry primarily by buying into nonpetroleum companies that own sizable stakes in oil companies as well as by buying into nonpetroleum companies that in turn channel Kuwaiti investments into oil companies.

One of its major outlets for downstream investments in Spain is the Barcelona-based paper manufacturer Torras Hostench, in which

Kuwait has a 45 percent holding as well as a sizable management presence. In 1987, Torras Hostench channeled Kuwaiti investment funds into the Spanish refiner and fertilizer manufacturer Unión Explosivos Río Tinto (ERT), the country's largest private sector chemical group, in which Kuwait now indirectly holds a 51 percent share. ERT owns the 80,000 b/d Huelva refinery, in which Kuwait is interested. ERT is also Spain's largest importer of Kuwaiti crude.[8]

Kuwait's interest in the Spanish market stems from the fact that gasoline consumption in Spain is projected to rise by 15 percent over the next four years. By 1992, of the projected 8,500 to 10,000 retail outlets in Spain, it is estimated that 800 to 1,500 will be owned by foreign companies. Since under Spanish regulations foreign companies are required to set up their own retail networks at the outset, the acquisition of a refinery would be only the first step in a much larger expansion Kuwait may have to undertake.[9]

In an effort to strengthen its position in the Asian market, Kuwait has contemplated buying into the Philippines National Oil Company (PNOC). While the Aquino government has considered the sale of 40 percent of PNOC, which owns a 150,000 b/d refinery and controls one-third of the country's retail gasoline market, there is significant domestic opposition to such a move.[10] But the prime target in Asia is Japan, whose government has thus far effectively barred any investment.

Another market, which Kuwait has tried to penetrate in the past and to which it may return, is the United States. Although Kuwait still prefers the European market because of its proximity, the possibility of owning refining and/or marketing properties in the United States is attractive. Kuwait has bid unsuccessfully for Northeast Petroleum Industries, the Boston-based New England marketing arm of the once bankrupt Charter Oil Company, a Gulf Oil/Chevron refinery in Louisiana, and the 174,000 b/d Gulf Oil/Chevron refinery in Philadelphia. It was also interested in the Texaco assets, which were acquired by Saudi Arabia. Kuwait has not established a foothold in the United States primarily because it is unwilling to offer premium prices for assets there.

Over 80 percent of Kuwait's oil exports are now sold as refined products. Worldwide product sales now total about 700,000 b/d, of which close to 500,000 b/d are exported from Kuwait. Africa consumes about 5 percent of Kuwait's product exports; South Asia, 20 percent; the Far East, 18 percent; Japan and the Middle East, 8 percent each; the United States, 2 percent; and Europe, the lion's share at 39 percent. By 1990, Kuwait may be selling nearly all its oil in the form of products. Kuwait's three domestic refineries, which have recently undergone considerable upgrading and expansion, now have capacity of close to 750,000 b/d.[11] With its access to numerous domestic and foreign refining outlets and such high potential output levels, it may have to find ways to circumvent its OPEC quota.

The efficiency of Kuwait's network of European service stations, with the exception of Scandinavia, is characterized by operating and distribution costs higher than the national averages of the countries in which they operate. In the United Kingdom and Italian markets, the increasingly efficient outlets of major companies and supermarkets may require Kuwait to invest substantial amounts of capital to insure the profitability of its investment. At the same time, Kuwait has attracted customers with its highly innovative marketing strategy. Kuwait's European stations, which offer a wide range of special services, are attractive and spacious. Kuwait introduced lead-free gasoline in Europe and many of its outlets have free self-service machines to change oil. The Q8 brand name has been well advertised, so its public recognition is high.[12]

Venezuela

While Kuwait chose to acquire full ownership of its downstream assets, Venezuela favors joint ventures in overseas downstream markets. The Venezuelan position is that joint ownership reduces political risk while it allows the sharing of market and economic risks as well as investment costs. The one exception to this rule was Venezuela's decision in late 1988 to acquire the remaining 50 percent of Champlin Petroleum's assets, which it now owns 100 percent.

As in the case of Kuwait, Venezuela's first joint venture was concluded in Europe in 1983, when it acquired a half-share in a West German refinery complex owned by Veba Öl, with a processing capacity of 210,000 b/d, for about $55 million. Initially, Venezuela supplied 450,000 b/d to the joint venture company known as Ruhr Öl, but in 1986 the joint venture agreement was expanded to include two more refineries in southern Germany and other facilities (i.e., a 50 percent share in Veba Öl's participation in the South Europe oil pipeline, which runs from southern France across Europe, and Veba Öl's share in the Transalpine pipeline, running from Trieste to its Neustadt refinery). Venezuela paid close to $55.4 million for this latest joint venture agreement with Veba Öl, which supplies 145,000 b/d of Venezuelan crude to West Germany, with the products sold through Veba Öl's distribution network.[13] Venezuela now controls 50 percent of Veba Öl. The Veba Öl joint venture provides a steady outlet for Venezuela's medium and light crudes and provides it with above average earnings, due in part to high German product prices. Venezuela receives a 50 percent share of all profits on product sales.

In 1986, Venezuela continued to expand its presence in the European market with its acquisition of 50 percent of Axel Johnson Company's subsidiary A.B. Nynas of Sweden for $25 million, to which it agreed to supply up to 40,000 b/d of naphthenic crude for lubricants and asphalt production. Nynas has two plants in Sweden and one in Belgium and ranks second to Shell in European sales of naphthenic lubricants.[14]

In the Caribbean Venezuela signed a five-year lease, paying $11 million per year, for the 320,000 b/d Curaçao refinery in the Netherlands Antilles in October 1985. The lease was extended in advance in November 1987 for an additional five-year period until 1994, with payments rising to $15 million per year.[15] Venezuela ships close to 170,000 b/d to the plant, which it agreed to lease to prevent economic and political dislocation when Shell decided to leave. This arrangement has given Venezuela the capability to increase its exports of asphalt and aromatics and provides substantial crude and products storage capacity and access to deepwater port facilities capable of accommodating VLCCs.

In September 1986, the U.S. market became another target in Venezuela's downstream joint venture/acquisition strategy. An agreement was reached to pay $290 million for a 50 percent interest in Citgo, the U.S. refiner and marketer which is a subsidiary of Southland Corporation.[16] Citgo is one of the largest independent refined products wholesalers in the United States, with volume of about 6 billion gallons per year of gasoline and other products sold through 4,000 Citgo-branded retail outlets, 3,500 Southland retail outlets, and airlines and other wholesale customers. This joint venture gave Venezuela half-ownership of Citgo's assets, including a 320,000 b/d refinery at Lake Charles, Louisiana (the eighth largest refinery in the United States); a 6 billion gallon per year wholesale marketing business; a lubricants blending and packaging plant (the sixth largest in the United States); 41 product terminals (5 of which are only partially owned by Citgo), providing 17 million barrels of storage; and varying interests in almost 16,000 miles of crude gathering and refined products pipelines. It also gave Venezuela a portion of the Cit-Con lubricants plant (the seventh largest in the United States) in Lake Charles, owned 65 percent by Citgo and 35 percent by Conoco. Venezuela agreed to supply up to 200,000 b/d of crude over a twenty-year period.[17]

In March 1987, Venezuela entered into another fifty-fifty joint venture with a U.S. company when it agreed to pay $93 million and to supply 140,000 b/d, with an option to provide 190,000 b/d, to Union Pacific Corporation's subsidiary Champlin Petroleum and its 160,000 b/d refinery in Corpus Christi, Texas.[18] In October 1988 it announced an agreement to buy all of the subsidiary, paying $75 million.[19] Champlin's distribution system supplies products through sixty-four marketing outlets in ten southern states.

By 1988, Venezuela supplied about 450,000 b/d of crude oil to its various joint venture arrangements. In addition it processed close to 170,000 b/d in the Curaçao refinery. Venezuela's goal is to have an outlet for 700,000 b/d of its crude oil in European and U.S. joint venture refineries. This amount represents 43 percent of Venezuela's estimated 1988 crude and product exports of 1.618 mb/d.[20]

Due to its OPEC quota, Venezuela was unable in 1988 to meet the supply needs of all its clients, including joint venture partners and third-party sales. Venezuela sold more oil than it actually produced. At times since 1987 it has had to purchase crude in international spot markets to supply part of its needs at the Curaçao facility and at the Citgo plant.[21] To ease the pressure, Venezuela intermittently cut back supplies to non–joint venture U.S. customers to minimum contract levels. Unless Venezuela's OPEC quota is raised significantly, increased purchases of crude overseas will be required not only to meet ongoing obligations but in the event that new joint ventures are concluded.

In December 1988 Venezuela signed a preliminary agreement to purchase 50 percent of Unocal's 147,000 b/d Chicago refinery, 14 terminals, more than 100 service stations, a lube oil blending plant, and contracts with 190 marketers. If the deal is completed, Venezuela will supply 135,000 b/d under a long-term contract.[22]

Table 9.1. Venezuela's Worldwide Refining Operations
(in thousands of barrels per day)

			1989	
	Capacity	*1988*	*Minimum*	*Maximum*
Domestic	1,224	798	—	—
Curaçao	320	170	160	185
Veba*	186	145	145	145
Nynas*	27	22	25	40
Citgo*	165	130	130	200
Champlin*	80†	140	140	160
Total	2,002	1,405	600	730

* Numbers refer to Venezuela's share of joint ventures. Venezuela acquired 100 percent of the Champlin refinery in October 1988; thus its crude supply obligations to Champlin increased in 1989.

† Based on Petroleum Finance Company estimates.

Venezuela exports nearly 600,000 b/d of products refined in its domestic plants while also shipping about 600,000 b/d of crude oil to its overseas joint-venture partners (in Europe and the United States) and the Curaçao refinery. It has more than 2 mb/d of refining capacity at its disposal and over 70 percent of its oil sales are already in the form of products from its domestic and foreign refineries. The U.S. market takes about 50 percent of its crude and product exports.

Table 9.1 shows the distribution of Venezuela's oil supplies among domestic and overseas joint-venture refineries and indicates Venezuela's 1986 exports by type of crude and product, including their volumes and destinations.

Libya

Libya's overseas downstream acquisitions are thus far limited to Italy and West Germany. In 1988, Libya established the Oil Investments International Company (OIIC) as a holding company controlled equally by the National Oil Corporation (NOC), the Libyan Arab Foreign Investment Company, and the Libyan Arab Foreign Bank. Capitalized with $500 million, the OIIC's purpose is to acquire downstream facilities overseas and to control all existing and future acquisitions.[23]

Existing acquisitions now controlled by OIIC include several downstream properties. In 1983, the Libyan Arab Foreign Bank took control of Tamoil Italia and used it as a vehicle to buy the refining and marketing assets of Amoco Italia, comprising 800 retail stations and a 105,000 b/d refinery at Cremona. It paid $263 million. In December 1985, the shareholder base of Tamoil was expanded so that it was owned 70 percent by the Libyan Arab Foreign Bank, 20 percent by Sasea (an Italian investor group), and 10 percent by Strand (an investor group led by Roger Tamarza).[24] The Libyan Arab Foreign Bank's 70 percent interest in Tamoil has now been taken over by OIIC.

Other existing acquisitions include the April 1987 purchase by Tamoil of the former Texaco marketing network in Italy. Consisting of some 770 service stations, 330 of which are wholly owned, the properties were sold by the Jacorossi group's Fintermica, Italy's biggest private sector energy conglomerate. Fintermica had bought these assets from Texaco in 1986. Tamoil Italia paid $100 million and increased its Italian service station holdings to 1,570.

In late 1987, Libya negotiated an agreement to purchase a 50 percent interest in a 90,000 b/d former Exxon West German refinery owned by Houston-based Coastal Corporation, which provides for the sale to be completed by 1993. Libya now supplies nearly 90,000 b/d to the refinery on a netback basis. However, in 1993 Libya may acquire as much as a 70 percent equity stake in the plant, in partial payment for releasing Coastal from a $50 million obligation under a 1980 concession agreement.[25] Any sale of equity Coastal works out with Libya is, however, likely to be reviewed by the U.S. government, and there is a possibility of the government taking some action. In September 1988, Tamoil Italia acquired 75 percent of Milan-based Vulcan Oil, a small gas oil and fuel oil marketer.[26] Libya now has foreign and domestic refining outlets for about 575,000 b/d, or 55 percent of its 1.037 mb/d quota.

Saudi Arabia

The most important step thus far taken in market reintegration is that of Saudi Arabia, which in one transaction became the producing country most committed to this process. Prior to the 1973 Arab oil embargo, Saudi Arabia considered buying into the U.S. refining industry. Political circumstances did not favor the implementation of this strategy and the Saudis instead undertook the construction of large joint-venture refineries on their own soil.

During 1985–86, Saudi Arabia was able to work out netback agreements and increased its crude oil sales through these agreements. After the 1986 December OPEC meeting, Saudi Arabia decided that downstream foreign integration would make it possible to move away from the role of OPEC's swing producer. The Saudi government moved cautiously at first in selecting potential downstream companies for acquisition, while a number of private Saudi investors were buying up attractive assets in the United States.

In June 1988 Saudi Arabia entered into a joint-venture agreement with Texaco for a 50 percent stake in its refining and marketing assets in twenty three eastern and Gulf Coast states and in the District of Columbia. A new joint venture company, Star Enterprise of Houston, was established. This acquisition provides an outlet for up to 600,000 b/d of Saudi crude, in effect giving the joint venture access to over 4 billion barrels of Saudi reserves over a twenty-year period at prevailing market prices. It is comprised of a stake in three Texaco refineries, including a 140,000 b/d plant in Delaware, a 250,000 b/d facility in Texas, and a 225,000 b/d plant in Louisiana.[27] This is 69.4 percent of Texaco's total U.S. refining capability of 886,000 b/d.[28] Other assets accruing to the joint venture include 49 terminals, 1,450 owned and leased service stations, and a branded distribution network of over 10,000 stations. The new venture began operations on 31 December 1988.

Saudi Arabia has become one of the largest gasoline retailers in the eastern United States, since Texaco's share of the gasoline market in this twenty-three-state region is 8 percent.[29] The three U.S. refineries, with 615,00 b/d of capacity, increase total Saudi refining capacity (domestic and foreign) to 2 mb/d. The venture also gives Saudi Arabia a major presence in U.S. products markets, where Texaco is the third largest supplier after Chevron and Shell. Texaco's overall share of the U.S. market is close to 9 percent.[30] For Saudi Arabia, the price was close to $1.5 billion, taking into account an $800 million payment (supposedly in cash but probably made up partially in oil), plus 75 percent of the initial crude and product inventories, or 30 million barrels (worth about $450 million at $15 per barrel) and 50 percent of the net working capital for the new venture. While the price was high, the Saudis probably calculated that it was worth the knowledge they would acquire about refined products markets in the United States. The acquisition also stabilized crude oil sales to the U.S. market, where Saudi exports have fluctuated within a very wide range since the late 1970s, from well over 1 mb/d in 1978 to under 200,000 b/d in 1983 and back to over 1 mb/d in 1988.

While some congressmen have expressed objections to the Texaco-Saudi arrangement on grounds of national security, neither the Justice nor Treasury departments has raised objections and, in fact, the venture has moved ahead without any problems.[31] The Saudis are said to be exploring other downstream options in Europe and they are being monitored closely by governments and by the oil industry.

Conclusion

The 1986 price collapse induced dramatic changes in the international petroleum industry that will shape its structure in the 1990s. The movement from setting prices administratively by exporters to market-based mechanisms marked a fundamental transformation in the way crude oil transactions are conducted. This change literally took decision-making control over pricing away from producers. It is likely to be a durable change in the way the oil business is conducted, even if oil markets grow tighter in the 1990s and if production again becomes concentrated in the Arabian Gulf. The growth of the futures markets, the so-called globalization of paper barrel trading, the diversification of supplies, and the movement downstream by key exporters make this prospect likely.

The overall impact of market reintegration itself, however, is much less clear. It now appears likely that the large-scale purchases of refining and distribution facilities in major markets by key producers is playing itself out. To be sure, additional acquisitions are likely by both Saudi Arabia and Kuwait. The large Japanese market, thus far closed to the exporters, is likely eventually to be open to some new ventures. But, even if another 2 mb/d of crude become tied up in the networks of internationally integrated national oil companies, that would mean only about one-fifth of current OPEC production would be involved in such arrangements. In short, this form of market integration has its limits.

The reasons for this are fairly convincing. Only a very few producers have the financial capacity or the human resources to acquire and manage an integrated oil company. Even Saudi Arabia appears to be unable to commit additional funds and personnel to these purposes despite government policy to do so, given the severe budgetary impact of lower prices and reduced production. Moreover, if oil prices begin to rise in the near future, there is bound to be a lower incentive for producers to seek out downstream marketing outlets. The value of refining and distribution facilities is bound to decline due to the fact that refining profits are almost certainly going to fall and may turn negative.

Therefore we can expect that the movement toward market reintegration will be limited and its impact on price stability for the industry is problematical. On the other hand, it has had a dramatic impact on relationships between state-owned producer companies and private sector enterprises. The joint venture operations of Saudi Arabia and Venezuela, especially, exemplify a much more pragmatic and balanced relationship between the two types of firms than existed in the 1970s environment of suspicion and lack of cooperation. Whether new pragmatism may effect additional types of cooperation, as in upstream exploration and production ventures, remains to be seen.

Notes

1. Petroleum Finance Company estimate for 1979; 1988 based on data from *Oil and Gas Journal*, 26 December 1988.
2. *Middle East Economic Survey*, 2 January 1989, p. D1.
3. *International Petroleum Finance*, 30 March 1984, p. 1.
4. *Journal of Commerce*, 29 October 1986.
5. *Petroleum Economist* (May 1987): 197; *Petroleum Intelligence Weekly*, 9 March 1987, p. 10.
6. *Oil and Gas Journal*, 11 May 1987, p. 18; *Platt's Oilgram News*, 28 May 1987, p. 1.
7. *Platt's Oilgram News*, 1 April 1987, p. 1; *Oil and Gas Journal*, 6 April 1987, p. 17.
8. *Wall Street Journal*, 4 January 1989, p. A3.
9. *Petroleum Intelligence Weekly*, 19 October 1987, p. 5; *Middle East Eco-*

nomic Digest, 31 October 1987, p. 18; *Middle East Economic Survey*, 2 November 1987, p. B1, 30 May 1988, p. B4.

10. *Platt's Oilgram News*, 10 June 1987, p. 4.

11. Ibid., 19 February 1987, p. 2, 9 June 1987, p. 4; *Petroleum Intelligence Weekly*, 5 October 1987, p. 2.

12. *Oil and Gas Journal*, 26 December 1988, p. 110.

13. *Platt's Oilgram News*, 11 June 1987, p. 1.

14. *Oil Daily*, 27 January 1986, 20 May 1987; *Petroleum Intelligence Weekly*, 12 May 1986, p. 3.

15. *Platt's Oilgram News*, 2 July 1986, p. 2.

16. *Petroleum Intelligence Weekly*, 2 November 1987, p. 7; *Oil and Gas Journal*, 16 November 1987, p. 22.

17. "The Southland Corporation—An In-Depth Look," Salomon Brothers, 16 March 1987.

18. Remarks by John P. Thompson, chairman of the board of the Southland Corporation, in Dallas, Texas, 15 September 1986.

19. *Platt's Oilgram News*, 13 March 1987, p. 3.

20. Ibid., 4 October 1988, p. 6.

21. Ibid., 29 December 1988, p. 1.

22. *Petroleum Intelligence Weekly*, 9 March 1987, p. 4.

23. *International Petroleum Finance*, 18 December 1988, p. 1.

24. *Middle East Economic Survey*, 5 September 1988, p. A3.

25. Ibid., 13 April 1987, p. A3.

26. Ibid., 5 September 1988, p. A4.

27. *Middle East Economic Digest*, 21 October 1988, p. 21.

28. *Oil and Gas Journal*, 21 November 1988, p. 26; *International Petroleum Finance*, 30 November 1988, p. 3; *Middle East Economic Survey*, 14 November 1988, p. D1.

29. *Petroleum Economist* (July 1988): 236.

30. Ibid.

31. Ibid.

JOHN H. LICHTBLAU

10. The Future of the World Oil Market

WAS the oil price collapse of 1986 inevitable or avoidable? The future of the world oil market depends to a considerable degree on the answer to this question. If we assume, as do many economists, that cartels are by their nature economic aberrations which sooner or later are superseded by market forces and if we look at OPEC as essentially a producers' cartel, 1986 may have signaled the beginning of OPEC's demise, notwithstanding repeated attempts to revive it. If, on the other hand, we assume that the governments of a small group of producer countries which collectively control the great bulk of world oil exports and whose production cost is at the very low end of the global range can theoretically maintain indefinitely some form of price control over their export commodity, the price collapse of 1986 may be viewed as the aberration and the subsequent attempt to revive the organization as a sign of its enduring vitality. Each answer leads to a different future price path and, hence, to a different supply and demand forecast. If the cartel is on its way out and the free market takes over, prices will obviously be substantially lower in the short and medium term than if OPEC were to retain (or regain) effective influence over price determination. Prior to 1986 the continued-cartel case, albeit at a declining price level, was considered the more likely by most forecasters. Since then the survivability of OPEC as an effective producers' cartel has been widely questioned, though generally not dismissed.

Actually, the 1986 price collapse was made inevitable only by OPEC's own actions. To be sure, the relentless dual pressure on OPEC since 1981 from declining world consumption and rising non-OPEC production severely tested its structure and posture. However, had the cartel at its December 1985 meeting gone back to its existing 16 mb/d production quota and had its members set their individual quotas in recognition of the fact that Saudi Arabia had irrevocably ended its role as the group's swing producer, prices would probably have stayed well above $20 per barrel and all cartel members would have earned substantially higher revenues in the 1986 to 1990 period than they actually did. As Dr. Chalabi points out in the introductory paper of this compendium, what OPEC did instead was to abandon its quota restrictions altogether, adopt a vague, undefined policy of obtaining a "fair market share" of world oil exports, and count on non-OPEC exporters to support the cartel by reducing production. In Dr. Chalabi's words, this led to "a state of absolute competition, not only between OPEC and non-OPEC countries, but also among its members." Thus, the 1986 price collapse was largely self-inflicted by OPEC and not an inevitable consequence of dynamic market forces or other factors beyond OPEC's control. In fact, one major extraneous factor which later hurt OPEC worked then in its favor: The Iran-Iraq war curtailed the output of both these major exporters for nearly eight years, starting in September 1980, and eliminated all of their considerable spare producing capacity.

How has the market reacted so far to the price collapse? World oil demand, which had declined more or less steadily from its 1979 peak through 1985, has reversed its direction in 1986. It has grown in every year since then. In 1990 demand in what used to be called the non-Communist world (NCW) market (excluding the Soviet bloc and the People's Republic of China) will be about 6 mb/d, or about 13 percent above the 1985 level, according to the International Energy Agency (IEA).[1] This represents an annual growth rate of 2.5 percent for the period 1985 to 1990. Total non-OPEC production in the NCW, which had risen steadily since the mid-1970s, has not shown a reversal following the price collapse. However, U.S. production, the world's second largest, did change its direction from flat in the first half of the 1980s to distinctly downward since 1986. In the first half of 1990 it was about 1.9 mb/d below the 1985 level. On the other hand, non-OPEC production outside the United States has risen almost as much since 1985 as it would have if there had been no price break.

While the increase in world demand and the decrease in U.S. supplies since 1986 are not insignificant, they reflect a very low price elasticity, considering that the 1986 price decline amounted to almost 50 percent and that in 1989 the annual world oil price was still one-third below the 1985 level in nominal dollars and about 50 percent in real dollars.

Where do we go from here? Let us limit our forecast to the medium term, that is, the five-year period to 1995, since it is less likely to contain market-impacting fundamental geological and technological changes than would a longer period. What price can we expect during this period? In a completely unconstrained market (a hypothetical case), the price would of course be a function of the interaction of supply and demand. In a partial cartel market (a realistic case), it reflects the policy of the cartel and its ability to enforce the policy.

The OPEC cartel had maintained a price from the early 1970s to

for the next few years OPEC will still have to cope with substantial excess producing capacity. From 1973 through 1981 the cartel had virtually no available excess capacity, nor did any other oil exporter. This was the key factor in OPEC's pricing power during that period. Hence, OPEC will not attempt to raise prices "excessively" in the next few years. The reason is the publicly acknowledged recognition by OPEC's principal producers, whose current or potential reserves/ production ratios are over one hundred years, that their long-term interests are better served by maintenance and eventual expansion of market shares than by price maximization at the risk of permanently losing market shares. Having had a dramatic demonstration in the 1981 to 1985 period of how much and how quickly market share can be lost when prices remain excessive, and having seen in the four yours since the big price break how difficult it is to regain the losses, OPEC's pricing policy in the first half of the 1990s will not be a repetition of the 1970s almost regardless of commercial circumstances. We can therefore expect an irregular but persistent moderate upward trend in real prices from next year on, despite the cartel's continuing surplus and continuing ability to meet all demand requirements at substantially lower prices.

What market trends to 1995 can we expect under this price scenario? In broad terms, a modest increase in world oil demand, say 1.5 to 2 percent per year, which would be less than the 2.5 percent growth rate of the 1985 to 1990 period; a leveling off in non-OPEC supplies, including Soviet and Chinese production, from 1990 on; and a steady increase in OPEC's world market share in the 1990s, with the bulk of the increase coming from the Middle East.

Among major markets, Europe will have little increase in oil demand because of the continuing inroads of other fuels (nuclear power and natural gas) into the stationary oil market. Japan, which had experienced almost steady annual declines from 1979 to 1985, registered an average annual growth rate of about 3.5 percent in the four following years and will probably continue to grow at 1.5 to 2 percent annually. The U.S. market, which is slightly larger than Europe and Japan combined, can be expected to grow at about 1 percent annually, increasing its volume by 1 to 1.2 mb/d between 1989 and 1995. Among major products the fastest growth in the United States will be in middle distillates. But U.S. residual fuel oil demand will also rise again from about 1992 on, primarily due to increased demand by electric utilities as the construction phase of nuclear power plants ends in the United States. The developing countries will register the most rapid growth rate, perhaps 3 to 3.5 percent annually, from their current (1990) consumption of 16 mb/d. In all major industrial countries other than the United States, residual fuel oil demand will continue to decline while light products demand will rise. In the developing countries, both light and heavy oil products will register a growth. But fuel substitution will be at work there, too, so that demand will grow much faster for light products than for heavy fuel oil in these countries.

On the supply side, non-OPEC NCW production was about 0.9 mb/d higher in 1989 than in 1985, the year before the price break. This is a very modest increase compared to those registered regularly in the pre-1986 period. The deceleration is due almost entirely to the aforementioned decline of 1.5 mb/d in U.S. production. For the rest of non-OPEC production the growth since 1986 has been almost as high as it would have been if prices had remained at their 1985 level.

The reason for the U.S. decline lies principally in the immediate sharp reduction in drilling activities following the price drop. In 1986 the number of active drilling rigs declined by 51 percent. By 1989 it had dropped by another 10 percent to its postwar low. Since the United States has by far the lowest output per well of any major oil producer, its production level is much more sensitive to the number of wells drilled at any given moment than almost any other producer. Outside the United States and OPEC the number of drilling rigs dropped by only 25 percent between 1985 and 1989. There are several reasons for the much smaller drop in drilling activities in these areas as well as the continuing increase in production. First, for most fields coming on stream in the 1986 to 1989 period, the investment decisions and most actual expenditures had been made several years earlier. Second, a number of countries (but not the United States) offset part of the 1986 price decline by reducing government taxes and/or royalties. Finally, finding and production costs in most countries are substantially lower than in the United States. Hence in those countries exploration and development activities did not have to be curtailed as much as in the United States nor did any flowing production have to be shut in at post-1985 prices. Other factors affecting exploration and development activities, both in the United States and abroad, are significant recent improvements in technology and inventive cost reductions implemented in the wake of the price break.

The general perception that real prices will rise slightly during the 1990s spurred a modest increase in global drilling activities in 1988–89 outside the United States and OPEC. This is likely to continue into the 1990s. By 1995 non-OPEC production ex United States may be 1 to 1.5 mb/d above the 1989 level, a 5 to 10 percent growth. U.S. production, on the other hand, will continue to decline. In fact, after 1989 the decline can be expected to accelerate because Alaskan production, which peaked in 1988 at 2 mb/d, has since entered its long-term decline phase. A recent decision by the U.S. administration to put much of the geologically promising Outer Continental Shelf areas at the East and West Coasts off limits for oil and gas drilling until at least the year 2000 as an environmental protection measure as well as the continued prohibition to drill in the Arctic National Wildlife Refuge (ANWR) will accelerate this decline. By 1995 U.S. production will have dropped by 1.2 to 1.4 mb/d from its 1989 level. This could offset most of the increase in other non-OPEC production during this period.

In arriving at a future balance of NCW oil supply and demand we must consider one other factor: net imports from the Soviet bloc and People's Republic of China. In 1985 net Soviet bloc exports amounted to 1.4 mb/d and exports from China to 733,000 b/d. By 1989 net Soviet bloc exports amounted to 1.5 mb/d (a 0.3 mb/d drop from the 1988 level) while Chinese net exports had dropped to about 400,000 b/d. The political and economic changes in the Soviet Union and eastern Europe are of such magnitude and rapidity and are so unprecedented that it is difficult at this time to project their energy requirements and export capability by 1995. We do know that the Soviet Union, the world's largest oil producer, has seen its production decline by 2.4 percent in 1989, from 12.8 mb/d to 12.5 mb/d; that the decline was progressive throughout the year; and that production is expected to decline further to 11.5 mb/d in 1990 and that Soviet exports to Western markets are reflecting the decline in production. But we also know that the Soviet Union wants to maintain its oil exports to the West because the West is its principal source of hard currency earnings, and we can assume that oil exports to eastern Europe, which in the past almost equaled those to the West, are likely to be curtailed in the future, both because the Soviet Union wants to sell more in the West and because the economic upheavals in most eastern European countries are likely to temporarily curb their energy requirements or their ability to pay for them. Let us assume, then, a net Soviet/Chinese oil export level of 1.4 mb/d by 1995, or 500,000 b/d less than in 1989.

We can now put together the NCW supply and demand balance for 1995. Demand, based on our annual growth rate of 1.5 to 2 percent, will be 57.5 to 59 mb/d in 1995. Non-OPEC supplies from all sources (including net exports from the Soviet Union, eastern Europe, and China) should be approximately 28 to 29 mb/d. Thus, depending on how we combine the upper and lower ends of the supply and demand ranges, OPEC, as the swing producer, would have to provide about 28.5 to 31 mb/d of liquids to balance supply and demand. Since the group's members can be expected to produce at least 2.5 mb/d of natural gas liquids by then, compared to 2 mb/d this year, its required crude oil production would range from just under 26 to slightly above 28 mb/d. The lower end of the range could be supplied from current OPEC capacity, which was about 28 mb/d in the first half of 1990, but it would require the organization to operate at 94 to 95 percent of capacity, a utilization rate reminiscent of the 1970s. The upper end of the required production range could of course currently not be met. Thus, the trend of world oil prices in the next five or six years will be strongly influenced by OPEC's willingness and ability to expand its productive capacity. We know that most of the countries able to increase their producing capacity are currently either actively engaged in doing so (Saudi Arabia is the prime but by no means only example) or are preparing plans for expansion. There is therefore no question that by 1995 the group's collective producing capacity will be significantly larger than in 1990. There is a question, however, whether the capacity increase will be big enough to accommodate the projected production increase and still leave enough spare capacity to keep the supply system flexible. In the 1970s this was generally not the case. In the 1980s it was excessively the case. In both decades insufficient and excessive spare capacity, respectively, were major factors in determining prices.

There is no indication that OPEC members, collectively or individually, intend to keep capacity expansion below requirements as a price-raising device. However, expansion costs money, and to the extent to which the money would have to come from government funds it may be given a lower priority than other state investments. Private foreign investments are of course available but only under the right economic and political conditions. These do not exist in all OPEC countries. There is therefore a risk of delays.

Current forecasts converge around an OPEC capacity level of 33 to 34 mb/d by 1995. If this level is reached, OPEC's spare capacity will be quite adequate and the intended real price increase during this period would require the five or six OPEC members with excess capacity to maintain production discipline by means of quotas or target prices throughout the period. If, on the other hand, capacity grows at a substantially slower rate, say, to only 30 mb/d by 1995, market forces are likely to take over and drive real prices up, almost regardless of OPEC's production policy. Thus, the rate of OPEC's capacity additions in the next five or six years will be a major determinant in how fast world oil prices will rise.

But whatever the combination of price, volume, and capacity assumptions in the various current forecasts, nearly all project a substantial increase in OPEC's export revenue in real dollars between 1990 and 1995. Lacking a world recession, this is a very plausible view.

Notes

1. International Energy Agency, *Oil Market Report* (June 1990): Table 1.
2. U.S. Department of Energy, *Annual Energy Outlook 1990* (Washington, D.C.: Energy Information Administration).

11. Epilogue: The Persian Gulf Crisis and the Oil Shock of 1990–1991

EDITOR'S NOTE: The Persian Gulf crisis and its related fourth oil price shock occurred as this book began the production process. To take account of these important events we asked three of the contributors to provide their preliminary analysis in this Epilogue.

Fadhil J. Al-Chalabi, for many years OPEC's Deputy Secretary General and now the Executive Director of the Institute of Global Energy Studies in London, provides a detailed account of the oil market background of Iraq's invasion of Kuwait. He notes that Kuwait's overproduction, an alleged reason for Iraq's military action, was not in itself sufficient explanation for the decline in oil prices in the spring of 1990 that hit Iraq especially hard. Rather, there was general OPEC excess production following a winter of high oil demand. Whatever its perception was, Iraq, which was unable to increase its production to gain more revenue in the short run, was clearly seeking a higher oil price to meet its financial needs than the $21 per barrel agreed upon at OPEC's July meeting.

Chalabi goes on to point out that the oil price escalation unleashed by Iraq's invasion of Kuwait and the subsequent U.N. trade embargo was driven primarily by the fear of war, not by a shortage of oil, since Saudi Arabia, Venezuela, and other OPEC countries increased their production to make up for the market shortfall. Whether the crisis is resolved by a military or a peaceful solution, Chalabi predicts a sharp oil price decline and an oversupply of oil when Iraqi and Kuwaiti capacity again re-enters the world market. Moreover, he foresees important structural changes in the world market after the crisis as the major oil consuming and importing countries try to reduce their dependence on oil from the Gulf. OPEC, he suggests, will no longer be able to have an important influence on the oil market unless it restructures itself to reflect more accurately the critical role of the core countries with the largest reserves and production capacities.

Edward N. Krapels, an oil consultant, offers a more detailed account of the oil price volatility in the autumn of 1990 which he sees as driven by concerns about adequacy of supply, adequacy of refining capacity, and fear of war and damage to other Gulf countries' oil facilities. He predicts an uneasy peace after the crisis, which will encourage governments to intervene more in the markets with energy security programs such as subsidies for oil production and gasoline taxes to reduce oil consumption. He seems pessimistic about prospects for OPEC cohesion and agrees with Chalabi on the likelihood of an oversupply of oil and falling prices after the crisis.

John H. Lichtblau, a prominent industry analyst, also foresees substantial overproduction of oil and a likely price crash when the war is over unless OPEC moves quickly to re-establish production quotas. In his view, and here he differs with Chalabi and Krapels, there are strong incentives for OPEC to regroup, prevent a decrease in prices, and even raise prices over the period 1991–95. However, non-OPEC production could increase as a result of higher oil prices during the crisis and world oil consumption will likely grow more slowly, which will cause problems for OPEC as the world's swing producer.

Background of the Gulf Crisis and Consequences for OPEC and for the Oil Industry

Fadhil J. Al-Chalabi

All was quiet in OPEC and the world oil industry during 1989 and the first quarter of 1990 before the clouds gathered to produce the gravest crisis in the history of the international oil industry. During that period the call on OPEC oil continued to grow, albeit at a slower rate than the average growth rate since 1986. Prices firmed up around the $18 per barrel target, and in some cases surpassed it. This good health of the oil industry was mainly due to the fact that world demand for oil continued to increase, whereas the world supply of oil outside OPEC did not show any sign of increase. On the average, the small increases that were noticed in some new areas in developing countries were generally offset by decreases in production in the older, high-cost producing areas. OPEC was producing well over its quota of about 21 mb/d yet prices were not showing signs of great weakness.

However, this calm situation in the international oil industry was short-lived as it preceded the great storm that shattered the entire Gulf and damaged intra-OPEC relations. The resulting short-term uncertainties are enormous. The far-reaching impact of the storm on the structure of the international oil industry, especially in the long-term, cannot at this writing be predicted.

The Gulf crisis started following a substantial decline in oil prices in the spring of 1990, when world demand fell sharply as a result of natural seasonal movement. The very high level of already accumulated oil stock build-up added another source of pressure on a price structure that was already weakened. With the summer approaching, the price decline became alarming, when prices went down to $15 per barrel and even lower. Consequently, oil revenues fell sharply, especially for countries that were producing at maximum capacity (the majority of OPEC), and thus were unable to compensate for the fall in prices by an increase in production. Of these countries Iraq was among the heavy losers.

Iraq suffered from a sharp decline in production capacity as a result of the lack of maintenance of the oil fields during the Iraq-Iran war, when Iraqi production had to be reduced from about 4 mb/d prior to the war to about 1.5 mb/d due to constraints on export outlets caused by the war. Toward the end of the war, and following the opening of the Saudi pipelines, Iraq's production had reached a maximum 3 mb/d. Being unable to maximize its revenues through higher volume, Iraq's only option was to raise prices. To a very great extent, this situation was behind the tragic events that resulted in the Iraqi invasion of Kuwait. Iraq was accusing Kuwait and also the UAE of overproducing by exceeding their OPEC quotas and flooding the market with such an amount of oil that it created an oversupply which caused the price fall.

In reality, however, the Kuwaiti overproduction (which amounted only to 380,000 b/d, against a production quota of 1.5 mb/d) could not by itself have caused a price decline of this magnitude ($6 to $7/barrel). OPEC's production over the previous months (over 23 mb/d) was very high, reflecting the then higher demand. In the spring of 1990 when demand started to fall in accordance with the seasonal pattern, OPEC as a whole did not adjust its production downward to maintain the price, but continued producing at the winter levels. Had OPEC wanted to maintain high oil prices, its total production should have been reduced to less than 21 mb/d. It was this production difference, amounting to about 2 mb/d and shared by all OPEC member countries, that was behind the price fall, and not the 380,000 b/d Kuwaiti excess production.

The next phase of the Gulf drama unfolded as Iraq formally requested that the OPEC target price be increased to $25 per barrel from its former level of $18 per barrel. For Iraq, with huge proven reserves, seeking wide, long-term market outlets, a $7 per barrel price hike could run against its long-term interests. However, given its grave financial problems, Iraq sided with all those oil-producing countries inside and outside OPEC which have very limited production capacities and for which only higher prices can maximize income or produce enough profits to allow their continued investment in the industry.

Conversely, for countries like Saudi Arabia and Kuwait (and, for that matter, Iraq and Iran also), such an increase could have damaging effects on the long-term demand for their oil, as $25 per barrel could simply reverse the rising trend in demand for OPEC oil that has been initiated since 1986 by OPEC's lower price profile. They resisted the price increase, however, and an OPEC compromise was reached, at a price of $21 per barrel.

By the time the OPEC conference was held in late July, Iraq was already massing large numbers of troops on the Kuwaiti border. At the same time, oil market indicators were pointing to the fact that the new OPEC price could not be sustained unless a substantial reduction in OPEC production was made, a reduction that was most difficult to achieve. Hence prices in the market after the OPEC conference were falling noticeably short of the new target price of $21 per barrel. This state of price weakness was dramatically reversed following the Iraqi invasion of Kuwait on August 2nd, when prices began soaring to unheard-of levels. The subsequent resolutions taken by the U.N. Security Council imposing a total blockade on both Iraqi and Kuwaiti oil led to an acute market crisis. Brent blend oil reached as high as $40 per barrel, almost twice as much as the OPEC price, although subsequently later in the fall the Brent price and other crude oil prices declined.

This sharp rise in prices was not caused by any physical shortage of oil resulting from the crisis and the shutdown of oil production of both Iraq and Kuwait. The net exports of both these two countries, about 4 mb/d, was more or less made up by Saudi Arabia's increasing production to 6.5 mb/d and then to 8 mb/d, against its 5.5 mb/d quota, and also Venezuela's stepped-up production to 2.5 mb/d contrasted with its 1.9 mb/d quota. Other small increases were also achieved in Nigeria, Libya, and Iran. These production increases indicated that in price terms the market was in balance. The price convulsion therefore reflected the political uncertainties of the Gulf crisis and the possibility that war might break out to end it, and not market fundamentals or supply/demand balances. Even the existence of the very high level of stocks that were available at that time, representing 99 days of forward consumption in the OECD countries, a level much higher than the minimum required, was not able to alleviate the nervousness in the market that was causing the price increase. The IEA decided not to release strategic stocks because a serious supply shortfall in the market did not exist.

In this context of great uncertainty, prices fluctuated violently, reflecting the psychology and perception of the market. Hence, prices went sharply down when the prospect of peace dominated the media headlines and conversely prices went up when the prospect of war seemed inevitable.

The 1990 Gulf crisis, like the one in 1979 that occurred in the wake of the Iranian revolution, and like the less acute crisis of 1980 created by the Iran–Iraq war, suggests that Gulf oil, which is crucial for the world oil balance, is most vulnerable to political events in the area, which are external to the oil industry. In fact, over the last 17 years, since the Arab oil embargo of October 1973, some three to four shocks of varying degrees have taken place in the world oil market, all being caused by political factors, but all with far-reaching disruptive effects on the industry. These events have shown how much Middle Eastern geopolitics is prone to dramatic and often tragic events that in turn are the result of very serious endemic sociopolitical problems such as the Arab-Israeli conflict and the political and social instability in some of the region's countries. The political frustration, especially concerning the Palestinian question, or the social frustration resulting from unequal distribution of the huge oil wealth nationally and regionally, or the total lack of guarantees of democracy and human rights in the area, will continue to be the direct or indirect causes of unrest that have historically caused political upheavals. It is a widely held view in oil industry circles that, unless these political problems are solved, the area will continue to be the scene of unpredictable dramatic events and disruptions which can cause far-reaching damage to the oil industry.

The short-term post-crisis situation, or course, depends on how the crisis ends. A military solution could be massively destructive to the oil installations in the area and might therefore create a physical shortage of oil, sending prices as high as the $70 to $80 per barrel level, only to fall dramatically afterwards once all the installations were repaired and the flow of oil from the Gulf resumed. It would only be a matter of time until an actual market glut took place. A peaceful solution, on the other hand, would mean that the shut-in capacity of both Iraq and Kuwait would come quickly on stream again, adding to the market some 4 to 4.5 mb/d of oil supplies, with the immediate effect of a sharp decline in the price of oil unless OPEC can defend the price by reducing production. This can prove to be most difficult, given the great reconstruction work required after the crisis in both countries, and the very high cost of the military presence of the Allied Forces in the Gulf.

More important is the impact of the Gulf crisis on the future of the world oil industry and, more particularly, on OPEC. As was the case with the previous crises, the 1990–91 crisis could provoke a host of forces that are capable of radically changing the structure of the world oil industry. The first reaction of the consumers/importers is that they will try to reduce dependence on oil imported from the Gulf. As oil developments since 1986 show, the continued increase in world demand for OPEC oil is satisfied in the main by supplies from the Gulf simply because the non-Gulf oil-producing countries have little capacity to expand. If we take the United States as an obvious case, we see that that country's imports have increased from less than 5 mb/d in 1985 to over 8 mb/d in 1989. This is the result of the continued increase in consumption of oil products, especially gasoline, and decreasing domestic production, reflecting mainly but not exclusively the decline rate of that country's mature oil fields. In fact, OPEC low price policies were accelerating the pace of the two opposing trends (higher consumption and lower domestic production), resulting in widening the oil gap, to be filled by oil imports. This rising trend of oil imports in the United States is expected to continue with a greater dependence on Gulf oil, it being the only source of incremental oil that could meet the growing world oil demand. In fact, more than 80 percent of incremental oil supplies in the United States are imported from the Gulf.

Such a growing dependence on an inherently unstable area could provoke political concern among the consuming countries from the point of view of security of supply. This could lead to certain countermeasures that could decelerate this growing dependence. For this reason, the crisis could turn the rising trend of Gulf oil exports into a stagnating if not a falling trend. Consumers and importers could achieve this by encouraging investment in oil outside the Gulf by means of various policy tools, including putting pressure on the Gulf producers to adopt a higher price regime. This could make investments in non-Gulf oil more attractive. OPEC's now high-price profile could only help accelerate this trend of lesser dependence on the Gulf. Obviously, it is the non-Gulf oil-producing countries that would benefit from that trend to the detriment of the Gulf.

On this issue and, more specifically, on that of the security of supplies from the Gulf, it is high time that some kind of international dialogue be started. Producers should create such political and legal frameworks that would encourage greater international cooperation in which international companies could play a tremendous role. It is only through a greater degree of interdependence between the various groups that a more lasting market stability can be achieved. What is needed is more integration in the world oil industry in order to reduce the present industry fragmentation and hence remove an important source of instability in the market. The major international oil companies could be the instrument of change toward the new era. Companies should be guaranteed greater access to crude oil (through new arrangements of crude lifting other than the concession system). Against this, oil-producing countries should be given greater assistance for investment in the downstream operations in the consuming countries.

On the other hand, the crisis has shown that OPEC in its present structure can no longer have a significant role to play in influencing the oil market. Apart from the Gulf countries and, to a lesser extent, Venezuela, no other member countries of OPEC are capable of increasing production capacity. The OPEC meeting that was held at the end of August following the shutdown of oil production in both Iraq and Kuwait was nothing but a political cover for the production increases in Saudi Arabia, Abu Dhabi, and Venezuela to make up for the shortage, which could have been done anyway with or without OPEC. This means that OPEC as an effective tool in price setting has lost its importance, whereas the high-reserve countries, the OPEC founders, namely Iran, Iraq, Kuwait, Saudi Arabia, and Venezuela, plus Abu Dhabi, have gained importance as the only countries that can provide the additional oil required by world trade. This means that in order for OPEC to be more effective and relevant it has either to comprise only countries with abundant oil reserves and oil production capacity (so as to be able to face and influence the changing situation of the world oil market), or to maintain its present structure (but with the voting power weighted in accordance with the relative importance of member countries in terms of reserves and production capacity). If OPEC fails to achieve this, it will be an organization without any real effectiveness in shaping the world market. The latter would in this case depend on the action of those high-reserve countries, which could be achieved outside OPEC individually or in concert.

Another long-term development, a result of the crisis, is that OPEC may, no matter what its structure, evolve a decision-making process that is heavily influenced by the regional political balances. We have seen this from the crisis, which actually started even before the invasion of Kuwait when Iraq in its quest for higher prices to offset the fall in its income resulting from the decline in prices, was in reality backing this demand by massing troops on the border in such a way as to enable it to force Kuwait reluctantly to accept higher prices. This kind of gunboat diplomacy could not have worked without the military might that Iraq possessed. Never in the history of OPEC has military force been used to impose a viewpoint within the organization.

The Oil Shock of 1990 and the Return to Energy Security

Edward N. Krapels

When Iraq invaded Kuwait on August 2nd, the world oil market was just getting over a period during which crude oil supply had substantially exceeded demand. Oil prices, which had begun the year at $22 per barrel, had dropped in June to $15.50 per barrel (WTI, near month), much to the dismay of OPEC's "price hawks." As had happened so often in the preceding three years, Kuwait and the United Arab Emirates had blatantly exceeded their production quotas, and it appeared that the oil market of 1990 would be a repeat of the post-1986 era, with no one country able to dominate events.

As the summer of 1990 approached, however, the level of intra-OPEC tension these price gyrations created seemed higher than usual. In the middle of July, Iraqi Foreign Minister Tariq Aziz prominently accused Kuwait of over-producing and stealing Iraqi oil from the Rumaila field, and his statements were shortly followed by deployment of Iraqi troops along the Kuwaiti border.

At the OPEC meeting of July 27 to August 1, the crude reference price was raised to $21 per barrel, and Kuwait and the UAE gave assurances, once again, that they would abide by their quotas. Subsequent events indicate that the credibility of these assurances was in question, at least in Baghdad. For a variety of reasons that doubtless included grievances over Kuwait's cavalier disregard of the oil price preferences of Iraq (and other OPEC price hawks), Iraq invaded Kuwait on August 2, 1990.

The immediate trade embargo organized by the America-led United Nations coalition cut off the flow of Iraqi and Kuwaiti crude oil to the world market. Combined with the cessation of Kuwaiti production (and a reduction in UAE output planned prior to the invasion and soon reversed), this reduced world crude oil supply in early August by 4.15 mb/d.

The oil price response to this loss was immediate and dramatic. As shown in Figure 11.1, spot WTI crude oil prices escalated to over $40 per barrel in October.

Prices rose to these levels for a variety of reasons:

One was the fear that a military confrontation between Iraq and the United States coalition would affect oil production in Saudi

Figure 11.1. The effects of the Persian Gulf crisis on international oil prices. Abbreviation: bbl, barrel.

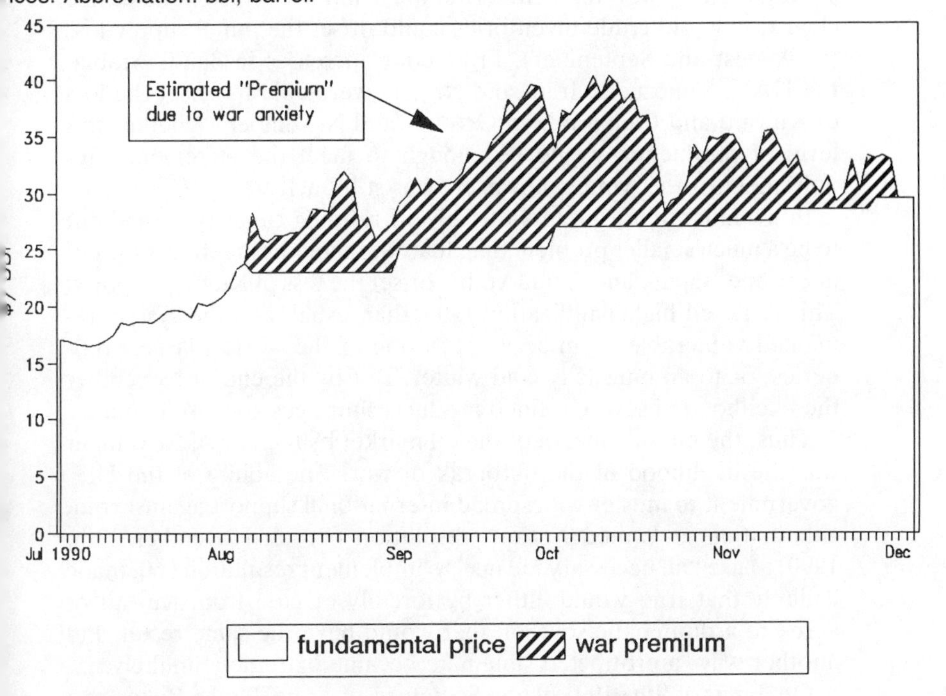

Arabia and the UAE. The price effect of this fear is depicted as the "war premium" in Figure 11.1.

A second factor was concern about the adequacy of oil supply, in terms of both production capacity and spare inventories. Would OPEC and other oil producers be able and willing to replace not only the loss of Iraqi and Kuwaiti crude but also provide the incremental supply that would be needed during the winter? That would require production everywhere to be stretched to maximum limits, limits that had not been tested in some countries in ten years.

A third factor behind the very sharp price increases was concern about the adequacy of global refinery capacity. Soon after the invasion, the operations of Kuwait's huge, sophisticated refineries were shut down, and with it exports of up to 500 thousand barrels per day of petroleum products. It was unclear early in the crisis whether other refineries in the complex network of the global industry would be able to pick up the slack.

By the end of November, each of these factors had been through several cycles of development. After much initial confusion, it became clear that world crude inventories could offset the initial supply loss (in August and September). Production increases in Saudi Arabia, the UAE, Venezuela, Iran, and Nigeria were able to offset the loss of Kuwaiti and Iraqi crude in October and November. Whether this level of production would be enough to meet the incremental requirement of the winter, however, was still unclear.

In a similar vein, the loss of Kuwaiti refining capacity turned out to be a much smaller problem than feared. Increases in refinery output in Europe, Japan, and Saudi Arabia offset the loss of Kuwaiti exports. This required higher utilization rates than usual, and the system remained vulnerable to an accident in one of the world's largest refineries, or to an unusually cold winter. But by the end of November the likelihood of severe refined products shortages appeared remote.

Thus, the major concern of the oil market by the end of November was the likelihood of the outbreak of war. The ability of the U.S. government to muster widespread international support against Iraq, culminating in United Nations resolution 678 on November 29th, 1990, to use "all necessary means" to implement resolution 660, made it likely that Iraq would either be forcibly ejected from Kuwait or agree to a diplomatic solution that would have the same result. Put another way, a protracted stalemate became extremely unlikely.

On January 9th, 1991, U.S. Secretary of State James Baker met in Geneva with Iraqi Foreign Minister Tariq Aziz in a "last ditch" effort to negotiate the withdrawal of Iraqi forces from Kuwait before the U.N. deadline of January 15. At the conclusion of the talks, when Baker announced to the press that his talks with Foreign Minister Aziz had produced no results, crude oil prices rose $6 within minutes of his opening statement. The reaction of the oil markets correctly indicated that the likelihood of war was now very high. One week later, on January 16th, American-led U.N. coalition forces began a punishing air assault on strategic targets in Iraq.

During the early hours of the coalition air strike, crude oil prices surged to $40 per barrel in the Far East. However, when the oil markets in London and New York opened the following morning, crude prices fell to the low 20s. The unanticipated price drop was largely a reaction to early reports from the Gulf suggesting that Iraqi resistance was limited and that coalition forces would quickly gain air superiority. The initial success of the air war diminished the perceived threat to Saudi production facilities. At the same time, the IEA announced a contingency plan that made available up to 2.0 mb/d of crude oil from strategic reserves (of which 1.125 mb/d would come from the U.S. SPR) and required the IEA member countries to provide an additional 0.5 mb/d through demand-restraint measures. As of this writing (early February) the impression that war with Iraq can be conducted without disrupting crude oil supplies has kept oil prices around $20 per barrel.

While it is uncertain how long the war will last, it is very likely that Iraqi forces will be forcibly removed from Kuwait and that 1991 will end with the return of independent Kuwait and the economic and military dismantling of Iraq. However, the end of the war will by no means bring stability to the region. The cost of this war, measured by both lives lost and dollars spent, will encourage governments to revitalize their energy security programs. Strategic reserves may be enlarged. Oil conservation and substitution programs are likely to receive more support, and world oil demand will grow less rapidly.

World oil supplies outside the Gulf are also likely to be aided by the continued perception of instability. Thus, non-OPEC output, including output in the United States, may receive direct or indirect assistance that otherwise would not have been available. In the United States, it is more likely that the outer continental shelf and the Arctic National Wildlife Refuge will be opened to exploration.

These demand and supply responses are likely to generate conditions that will cause the medium and long-term oil market outlooks to be substantially more bearish than they otherwise would have been. Demand for oil will be lower, and, once Kuwait and Iraq are back in the market, the supply of oil will be higher than most pre-Iraq invasion forecasts had reckoned.

Thus, the international oil market is likely to have substantially more spare oil production capacity after the Gulf war. International oil prices are likely to be lower. But, unlike the 1980s, low international oil prices will not necessarily mean low domestic oil prices. Governments may well subsidize domestic oil production and tax oil consumption in order to minimize the need for imports.

In other words, as a result of Iraq's invasion of Kuwait the 1990s are likely to see the return of government interest in oil, even in countries like the United States. Energy security has returned to the front burner. The oil world has changed course again.

The Gulf Crisis and the Future of the World Oil Market

John H. Lichtblau

The latest Gulf crisis brought about by Iraq's invasion of Kuwait and the subsequent UN-imposed embargo of all oil exports from both countries, is now entering its seventh month. It has brought drastic changes into the world oil market. Prices exploded instantly and remained high throughout the 5½ month period from the invasion of Kuwait to the start of the war on January 16th. The average price of West Texas Intermediate crude was over $30 during this period, compared with just under $20 in the first 7 months of 1990. However, on the first day of the war prices fell by a historic $12 and have remained in the $20 to $22 range through the first week of February (when this was written). The price crash was the reverse of all predictions of what would happen at the outbreak of war. It reflected a total turnaround in market perception from a preoccupation with potential war damage to Saudi oil facilities to a realization that the market was fundamentally balanced without any oil from Iraq or Kuwait, and that even a moderate loss of Saudi production would not create a shortage since it could be offset by drawing on the noncommercial strategic oil reserves held by the United States and other importing countries. Thus, in the first month of a major war in the Persian Gulf, with no immediate end in sight, crude oil prices were probably no higher than if the invasion of Kuwait had never taken place.

What will all this do to our price and volume projections for 1995, described in chapter 10? The answer depends of course on the conduct and duration of the war and the damages inflicted on the oil fields and installations. All assumptions on these points are of course heroic as of this writing. However, they need to be made for the purpose of our analysis. Let us assume that the war and the sanctions end sometime in 1991 and that Kuwaiti and Iraqi oil exports will have been physically restored to their prewar level within 6 to 8 months thereafter. Under this assumption, our projections to 1995 would remain generally valid, as described below.

Prices can be expected to decline to less than their current level when the war ends, but the likely inability of Iraq and, even more so, of Kuwait to export any oil in the early postwar period because of damage to their production, transportation, and loading facilities is likely to keep the decline from becoming a rout, as had been predicted before the start of the war when serious damage to oil installations was not viewed as likely.

OPEC has been in suspended animation since August 2nd because quota ceilings and floor prices have become irrelevant. However, the commitment of all members at OPEC's December meeting to return to their assigned OPEC quota when the crisis ends is not irrelevant in projecting postwar price trends. That decision reflects the economic and political self-interest of all OPEC members, namely, to reestablish a postwar production quota to support the agreed reference price. Future intra-OPEC discipline in this regard may be quite different from what it was in the recent pre-crisis period. One of the Iraqi dictator's prime accusations against Kuwait was that its consistent substantial exceedence of its assigned and agreed OPEC quota was a major factor in depressing world oil prices and was intended to weaken the Iraqi economy. It is unlikely that either Kuwait or the UAE, the other consistent quota overproducer, will return to this policy in the post-crisis period or that Saudi Arabia will adopt it. If Sadam Hussein is still in power, these countries will be afraid to do so, and if he has been replaced, they would want to help his successor to rebuild his country's shattered economy by supporting the agreed price structure. The likely better discipline among OPEC Middle East producers, together with the increase of OPEC's official minimum reference price from $18 to $21, under strong pressure from Iraq just five days before the invasion of Kuwait, should cause all Middle East producers to strive for a somewhat higher price in the 1991–95 period than was projected in the spring of 1990.

However, there are some offsets. One is that the sharply higher prices since the beginning of August have already led to additional investments in non-OPEC production that will therefore increase modestly, but measurably, from what it would have been otherwise. Another factor is that OPEC producers outside the Middle East which face no political or military risks in exceeding their quota may well do so in the post-crisis period. In Venezuela there is currently a public debate on this issue. Finally, the price shock of the crisis has caused world oil consumption to decline temporarily. Quite possibly it could be lower by 1995 than if there had been no price shock. The reduction would fall entirely on OPEC as the world's swing producer.

To sum up, the Persian Gulf crisis may cause world oil demand to be slightly lower, non-OPEC oil production slightly higher, and intra-OPEC price discipline somewhat better over the next five years than if the oil crisis of 1990–91 had never occurred.

The Contributors

WILFRID L. KOHL is director of the International Energy Program at the Johns Hopkins Foreign Policy Institute and research professor of international relations at SAIS. His earlier publications include *International Institutions for Energy Management* (Gower Press for the British Institute's Joint Energy Policy Programme, 1983) and, as editor and co-author, *After the Second Oil Crisis: Energy Policies in Europe, North America and Japan* (Lexington Books, 1982) and *Methanol as an Alternative Fuel Choice: An Assessment* (Johns Hopkins FPI, 1990). He taught previously at Columbia University and has served on the staffs of the Ford Foundation and the National Security Council.

MELVIN A. CONANT is president of the consulting firm of Conant and Associates, Ltd., which advises governments and industry on the political aspects of international energy questions. He edits *Geopolitics of Energy,* a monthly report, and is the author of several books on energy policy. He is also chairman of the advisory committee of the SAIS International Energy Program. In the mid-seventies he was assistant administrator for international affairs at the Federal Energy Administration. Prior to that he worked for a dozen years with Exxon Corporation.

FADHIL J. AL-CHALABI, currently executive director of the Institute for Global Energy Studies in London, was deputy secretary general of OPEC from 1978 to 1989 and acting secretary general for a number of those years. During this period he participated in key OPEC policy meetings and nearly all ministerial conferences. He was also editor in chief of *OPEC Review.* Dr. Chalabi is the author of *OPEC and the International Oil Industry: A Changing Structure* (Oxford University Press, 1980) and numerous articles on energy and oil economics.

HOSSEIN ASKARI is professor of international finance and director of the International Business Program at George Washington University. Previously he was an adviser to the executive director (for Saudi Arabia) at the International Monetary Fund and to the minister of finance of Saudi Arabia. He has taught at Tufts University, the University of Texas, and at Johns Hopkins SAIS. His latest book is *Saudi Arabia: Oil and the Search for Economic Development* (JAI Press, 1988).

EDWARD N. KRAPELS is president of Energy Security Analysis, a Washington, D.C., consulting firm. His previous publications include *Pricing Petroleum Products: Strategies of Eleven Industrial Countries* (McGraw-Hill, 1982) and *Oil Crisis Management: Strategic Stockpiling for International Security* (Johns Hopkins University Press, 1980). His chapter in this book is drawn from a forthcoming study, *The Commanding Heights of Oil: Control over the International Oil Market.*

DAVID H. VANCE is an energy economist in the Bureau of Intelligence and Research at the U.S. Department of State. He has also worked in the Office of Emergency Preparedness and the U.S. Treasury Department.

PHILIP K. VERLEGER, JR., is an energy economist and senior fellow at the Institute of International Economics. He is also a consultant with Charles River Associates and the author of the *CRA Petroleum Economics Monthly*. His publications include *Oil Markets in Turmoil* (Ballinger, 1982) and numerous articles on energy economics. He has served on the staffs of the Council of Economic Advisers and the U.S. Treasury Department. He has also taught at Yale University.

CAROL A. DAHL is associate professor of economics at Louisiana State University in Baton Rouge. After receiving her Ph.D. in economics from the University of Minnesota, she taught economics at Wayne State University, the University of Michigan, and the University of Wisconsin, Milwaukee. She has been a research fellow at M.I.T. Energy Laboratory and the Oxford Institute for Energy Studies. Her research has focused on energy demand analysis and the modeling of international energy markets.

DILLARD P. SPRIGGS is president of Petroleum Analysis, Ltd., a New York-based consulting firm specializing in petroleum economics and finance. He also edits the firm's industry report, *International Petroleum Finance,* which is published bimonthly and specializes in management strategies, earnings, and finances of oil companies.

EDWARD L. MORSE is managing director of the Petroleum Finance Company in Washington, D.C., and executive publisher of *Petroleum Intelligence Weekly* in New York. Previously, he was director of international affairs at the Phillips Petroleum Company and deputy assistant secretary of state for International Energy Policy. He also taught international relations for a number of years at Princeton University. Dr. Morse is the author of several studies in political economy, international politics, and foreign policy and of articles on OPEC and the world oil market.

JULIA NANAY is director of trade and advisory services at the Petroleum Finance Company. Previously she was vice-president of an independent oil company. She holds a master's degree from the Fletcher School of Law and Diplomacy.

JOHN H. LICHTBLAU is president of the Petroleum Industry Research Foundation in New York and a leading international expert on petroleum economics. He also chairs a consulting firm, Petroleum Industry Research Associates. He is the author of numerous articles and has been a frequent witness at congressional hearings on energy policy, as well as a frequent speaker at energy conferences in the United States and abroad.

Index

Designed by Sue Bishop
Composed by Easton Publishing Services, Inc.
in Linotron Times Roman.
Printed on 50-lb., BookText Natural
and bound in Roxite vellum cloth
by BookCrafters.